Dynamics Reported

Expositions in Dynamical Systems

Dynamical Systems are a rapidly developing field with a strong impact on applications. Dynamics Reported is a series of books of a new type. Its principal goal is to make available current topics, new ideas and techniques. Each volume contains about four or five articles of up to 60 pages. Great emphasis is put on an excellent presentation, well suited for advanced courses, seminars etc. such that the material becomes accessible to beginning graduate students. To explain the core of a new method contributions will treat *examples* rather than general theories, they will describe *typical results* rather than the most sophisticated ones. Theorems are accompanied by *carefully written proofs*. The presentation is as *self-contained* as possible.

Authors will receive 5 copies of the volume containing their contributions. These will be split among multiple authors.

Authors are encouraged to prepare their manuscripts in Plain TEXor LATEX. Detailed information and macro packages are available via the Managing Editors.

Manuscripts and correspondence should be addressed to the Managing Editors:

C.K.R.T. Jones
Division of Applied Mathematics
Brown University
Providence, Rhode Island 02912
USA
e-Mail: ckrtj@cfm.brown.edu

U. Kirchgraber
Mathematics
Swiss Federal Institute
of Technology (ETH)
CH-8092 Zürich, Switzerland
e-Mail: kirchgra@math.ethz.ch

H.O. Walther
Mathematics
Ludwig-Maximilians University
D–80333 Munich
Federal Republic of Germany
e-Mail: Hans-Otto.Walther
@mathematik.
uni-muenchen.dbp.de

C. K. R. T. Jones U. Kirchgraber H. O. Walther

(Managing Editors)

Dynamics Reported

Expositions in Dynamical Systems

New Series: Volume 3

With Contributions of
G. Fournier, I. Lasiecka, D. Lupo, Y. Nishiura,
M. Ramos, R. Triggiani, M. Willem

Springer-Verlag
Berlin Heidelberg New York
London Paris Tokyo
Hong Kong Barcelona
Budapest

ISBN-13:978-3-642-78236-7   e-ISBN-13:978-3-642-78234-3
DOI: 10.1007/978-3-642-78234-3

Library of Congress Cataloging-in-Publication Data
Dynamics reported: expositions in dynamical systems/C.K.R.T. Jones, U. Kirchgraber, H.O. Walther,
managing editors: with contributions of R. Fournier ... [et al.]. p. cm.
ISBN-13:978-3-642-78236-7

I. Differentiable dynamical systems. I. Kirchgraber, Urs, 1945– . II. Walther, Hans-Otto.
III. Bielawsi, R. QA614.8D96 1991 003′.85–dc20 91-23213 CIP

© Springer-Verlag Berlin Heidelberg 1994
Softcover reprint of the hardcover 1st edition 1994

41/3020-5 4 3 2 1 0 – Printed on acid-free paper

Preface

DYNAMICS REPORTED reports on recent developments in dynamical systems.

Dynamical systems of course originated from ordinary differential equations. Today, dynamical systems cover a much larger area, including dynamical processes described by functional and integral equations, by partial and stochastic differential equations, etc. Dynamical systems have involved remarkably in recent years. A wealth of new phenomena, new ideas and new techniques are proving to be of considerable interest to scientists in rather different fields. It is not surprising that thousands of publications on the theory itself and on its various applications are appearing

DYNAMICS REPORTED presents carefully written articles on major subjects in dynamical systems and their applications, addressed not only to specialists but also to a broader range of readers including graduate students. Topics are advanced, while detailed exposition of ideas, restriction to *typical* results – rather than the *most general* ones – and, last but not least, lucid proofs help to gain the utmost degree of clarity.

It is hoped, that *DYNAMICS REPORTED* will be useful for those entering the field and will stimulate an exchange of ideas among those working in dynamical systems

Summer 1991   Christopher K.R.T Jones
                                                   Urs Kirchgraber
                                                   Hans-Otto Walther

                                                   Managing Editors

Table of Contents

Recent advances in regularity of second-order hyperbolic mixed problems, and applications

I. Lasiecka, R. Triggiani

Limit Relative Category and Critical Point Theory[1]

G. Fournier[2]
D. Lupo[3]
M. Ramos[4]
M. Willem

1. Introduction

Let X be a Finsler manifold of class C^1 and let $f \in C^1(X, \mathbb{R})$. A *critical point* of f is a point $u \in X$ such that $df(u) = 0$. The corresponding value $f(u)$ is called a *critical value*. In many applications, the critical points of a functional are the weak solutions of a boundary value problem. For example, the critical points of

$$f(u) = \int_0^T [-\frac{1}{2}(Ju, u) - H(t, u)] \, dt \qquad (1)$$

on an appropriate space of T-periodic functions are the T-periodic solutions of the Hamiltonian system

$$Ju + \nabla H(t, u) = 0,$$

where ∇ denotes the gradient with respect to u.

The most obvious candidate for a critical value is the infimum of the functional

$$c = \inf_{u \in X} f(u),$$

when it is finite. In order to obtain other critical values, the minimax method was introduced by Lusternik and Schnirelman in 1930. The value c is then defined by

$$c = \inf_{A \in \mathcal{A}} \sup_{u \in A} f(u)$$

where $\mathcal{A}$ is a suitable class of subsets of X. When f is bounded from below, the set $\mathcal{A}$ can be defined by using an interesting topological invariant : the Lusternik-Schnirelman category.

This was done in 1930 by Lusternik and Schnirelman for compact manifolds. The

---

[1]Received: November 27, 1990
[2]G. Fournier, University of Sherbrooke, Sherbrooke, Canada.
[3]Università di Trieste, Trieste, Italy.
[4]CMAF/INIC, Lisbon, Portugal

theory was extended to Riemannian manifolds by J.T. Schwartz in 1964 and to Finsler manifolds by Palais in 1966 [Pa$_1$]. The main ingredients are a compactness condition, called *Palais-Smale condition*, and the notion of *pseudo-gradient flow* which generalizes to Finsler manifolds the notion of gradient flow.

On the other hand, the minimax method was applied in many situations where f is unbounded from below without using category. The most elementary example is the Ambrosetti-Rabinowitz Mountain Pass theorem (1973). But a year before, in a very interesting paper [Re], M. Reeken introduced a notion of relative category applicable to functionals which are unbounded from below. Unfortunately the paper was forgotten and the notion was rediscovered by Fournier and Willem ([FW$_1$]). The relative category is applicable to (1) only after a Liapunov-Schmidt reduction ([FW$_4$]) because the main part of the functional, the quadratic form

$$\int_0^T -(Ju, u)\, dt,$$

has infinitely many positive eigenvalues and infinitely many negative ones. The relative category was recently extended to strongly indefinite functionals by A. Szulkin [Sz$_1$] using a device of Benci and Rabinowitz ([BR]). Similar problems can also be studied by using the pseudo-index of Benci [Be]. For related problems with symmetry see for instance [CP].

In the present paper, we propose another extension of the relative category applicable to strongly indefinite functionals. As in [Li], we use a Galerkin argument and the compactness condition $(PS)^*$ introduced in [BB], [LL]. Our approach is quite elementary. For example, we obtain the saddle point theorem and the linking theorem for strongly indefinite functionals by using only the non retractability of the ball into the sphere in finite dimension.

Moreover, by using singular cohomology, we can consider these theorems on the product of a compact manifold with a Banach space.

The paper is organized as follows. In the second section we present a variant, due to Szulkin, of the relative category introduced in [Re$_1$] and [FW$_1$]. The third part is devoted to the relative cuplength which gives an estimate from below of the relative category. In some situations, it is easier to compute the relative cuplength. The fourth part is devoted to the limit relative category. In parts 5 and 6 we prove the corresponding deformation lemma and critical point theorems. Part 7 is devoted to a by now classical application to Hamiltonian systems motivated by [CZ]. Part 8 contains a perturbation theorem motivated by the results in [FW$_1$], [CLZ], [Fe$_1$], [Ta], [Ta$_1$] on the N-pendulum equation.

2. Relative Category

We recall some basic definitions and set up some terminology. By a *map* between topological spaces we mean a continuous function. Let (X, A) be a topological pair ; a *deformation* $h_t : A \to X$ is a map $h : [0, 1] \times A \to X$ such that $h_0(x) = x$ for every $x \in A$. The set A is *contractible* in X if there exists a deformation $h_t : A \to X$ such that $h_1(x) = h_1(y)$ for every $x, y \in A$.

Let A, B, Y be closed subsets of the topological space X. Then, by definition, $A <_Y B$

in X if $Y \subseteq A \cap B$ and if there exists a deformation $h_t : A \rightarrow X$ such that $h_1(A) \subseteq B$ and $h_t(Y) \subseteq Y$ for every $t \in [0, 1]$.

We now introduce the concept of relative category :

Definition 2.1. *Let $Y \subseteq A$ be closed subsets of a topological space X. The* relative category *of A in X relative to Y is the least integer n such that there exist $n + 1$ closed subsets $A_0, A_1, \ldots, A_n$ of X satisfying*

(a) $A = \bigcup_{i=0}^{n} A_i$;
(b) $A_1, \ldots, A_n$ are contractible in X ;
(c) $A_0 <_Y Y$ in X.

When no such integer exists, the category of A in X relative to Y is infinite. The relative category is denoted by $cat_{X,Y}(A)$.

When $Y = \phi$ the relative category $cat_{X,\phi}(A)$ is, by definition, equal to the Lusternik-Schnirelman category $cat_X(A)$. Considering $A_0 = Y$ in the above definition we see that $cat_{X,Y}(A) \leq cat_X(A)$ whenever $Y \subseteq A$.

We give some few examples :

Example 2.2.

a) Let $A := \{x \in \mathbb{R}^N : 1 \leq \| x \| \leq 2\}$, $N \geq 2$; denote by ∂A the boundary of A. Then $cat_{A,\partial A}(A) \leq 2$ (we shall see below that in fact equality holds).

b) If H is an Hilbert space and B is its unit ball then

$$cat_{B,\partial B}(B) = \begin{cases} 1 & \text{if } \dim E < \infty, \\ 0 & \text{if } \dim E = \infty. \end{cases}$$

c) Consider $S = \{(x, y) \in \mathbb{R}^2 : \| (x, y) - (0, 2) \| < 1\}$, $X = \mathbb{R}^2 \backslash S$, $Y = \mathbb{R} \times \{0\}$, $A = (Y \cup] -\infty, 0] \times [0, +\infty[) \cap X$, $B = (Y \cup [0, +\infty[\times [0, +\infty[) \cap X$. Then $cat_{X,Y}(A) = 0 = cat_{X,Y}(B)$ while $cat_{X,Y}(A \cup B) = 1$.

d) Consider $X = \mathbb{R}$, $Y = \{0\}$, $Y' = \{0, 1\}$, $Y'' = [0, 1]$. Then $Y' <_Y Y$ and $cat_{X,Y}(X) = 0$ but $cat_{X,Y'}(X) = 1$ and $Y \subset Y' \subset Y''$ but $cat_{X,Y}(X) = 0$, $cat_{X,Y'}(X) = 1$, $cat_{X,Y''}(X) = 0$.

Lemma 2.3. *Let A, B, C, Y be closed subsets of X with $Y \subseteq A \cap B \cap C$. If $A <_Y B$ and $B <_Y C$ in X then $A <_Y C$ in X.*

Proof: Assume $A <_Y B$ and $B <_Y C$ by means of deformations $h_t : A \rightarrow X$, $k_t : B \rightarrow X$ respectively. Then the map $k \star h : [0, 1] \times A \rightarrow X$ given by

$$(k \star h)(t, x) = \begin{cases} h(2t, x) & , \quad 0 \leq t \leq \frac{1}{2} \\ k(2t - 1, h(1, x)) & , \quad \frac{1}{2} \leq t \leq 1 \end{cases}$$

is a deformation such that $(k \star h)(1, A) \subseteq C$ and $(k \star h)(t, Y) \subseteq Y$ for every $t \in [0, 1]$. $\square$

Proposition 2.4. *Let A, B, Y be closed subsets of X with $Y \subseteq A \cap B$. The relative category satifies the following properties:*

(a) Normalization : $cat_{X,Y}(Y) = 0$;
(b) Subadditivity : $cat_{X,Y}(A \cup B) \leq cat_{X,Y}(A) + cat_X(B)$;
(c) Homotopy : if $A <_Y B$ then $cat_{X,Y}(A) \leq cat_{X,Y}(B)$;
(d) Monotonicity : if $A \subseteq B$, then $cat_{X,Y}(A) \leq cat_{X,Y}(B)$.

Proof: (a) follows from the trivial fact that $Y <_Y Y$ and (d) follows from (c). To prove (b), let $n = \mathrm{cat}_{X,Y}(A)$, $k = \mathrm{cat}_X(B)$, write $A = \bigcup_{i=0}^n A_i$ as in Definition 2.1, and $B = \bigcup_{i=1}^k B_i$ where each closed subset B_i is contractible in X. Then $A \cup B = A_0 \cup \left(\bigcup_{i=1}^n A_i \cup \bigcup_{i=1}^k B_i \right)$ where $A_0 <_Y Y$ so that $\mathrm{cat}_{X,Y}(A \cup B) \le n + k$.

As for (c), let $h_t : A \to X$ be such that $h_1(A) \subseteq B$ and $h_t(Y) \subseteq Y$. Writing $B = \bigcup_{i=0}^n B_i$ as in Definition 2.1, where $n = \mathrm{cat}_{X,Y}(B)$, and letting $A_i := h_1^{-1}(B_i)$ we have $A = \bigcup_{i=0}^n A_i$; clearly $A_0 <_Y B_0$, $A_i <_\phi B_i$ for $i = 1, \ldots, n$ and Lemma 2.3 implies that $\mathrm{cat}_{X,Y}(A) \le n$. $\qquad\square$

Remark 2.5. Together with A. Szulkin we noticed the following maximality property of the relative category : if $\varphi_{X,Y}$ is some function defined on the class of closed subsets of X containing Y satisfying (a), (b), (c) of Proposition 2.4, then $\varphi_{X,Y}(.) \le \mathrm{cat}_{X,Y}(.)$. Indeed, if $n = \mathrm{cat}_{X,Y}(A)$ and $A = \bigcup_{i=0}^n A_i$ according to the definition then

$$\varphi_{X,Y}(A) \le \varphi_{X,Y}(A_0) + \mathrm{cat}_X \left(\bigcup_{i=1}^n A_i \right)$$

$$\le \varphi_{X,Y}(Y) + \sum_{i=1}^n \mathrm{cat}_X(A_i) = n.$$

Proposition 2.6. *Let $Y \subseteq A$ be closed subsets of X, $Y' \subseteq A'$ be closed subsets of X'. Suppose there exist maps*

$$(X, A, Y) \xrightarrow{f} (X', A', Y') \xrightarrow{g} (X, A, Y)$$

and a deformation $h_t : X \to X$ such that $h_1 = g \circ f$, $h_t(Y) \subseteq Y$ for every $t \in [0, 1]$. Then

$$\mathrm{cat}_{X,Y}(A) \le \mathrm{cat}_{X',Y'}(A').$$

Proof: Write $A' = \bigcup_{i=0}^n A_i'$ according to definition 2.1 and set $A_i := f^{-1}(A_i')$, $i = 0, \ldots, n$. Then $A \subseteq \bigcup_{i=0}^n A_i$ and $Y \subseteq A_0$. Let $k_t : A_0' \to X'$ be a deformation such that $k_1(A_0') \subseteq Y'$, $k_t(Y') \subseteq Y'$, $0 \le t \le 1$; then the map $\bar{h}_t : A_0 \to X$ given by

$$\bar{h}(t, x) = \begin{cases} h(2t, x) & , \quad 0 \le t \le \frac{1}{2} \\ g(k(2t - 1, f(x))) & , \quad \frac{1}{2} \le t \le 1. \end{cases}$$

is well-defined and shows that $A_0 <_Y Y$. Similarly, each set A_i, $1 \le i \le n$, is contractible in X and $\mathrm{cat}_{X,Y}(A) \le n$. $\qquad\square$

We say the pairs (X, Y), (X', Y') have *the same homotopy type* if there exist maps $(X, Y) \xrightarrow{f} (X', Y') \xrightarrow{g} (X, Y)$ and deformations $h_t : X \to X$, $k_t : X' \to X'$ such that $h_1 = g \circ f$, $k_1 = f \circ g$, $h_t(Y) \subseteq Y$ and $h_t(Y') \subseteq Y'$.

If $X \subseteq X'$, by a *retraction* we mean a map $r : X' \to X$ such that $r(x) = x$ for every $x \in X$.

Corollary 2.7. *Let Y, Y' be closed subspaces of X and X' respectively.*

a) *If (X, Y) and (X', Y') have the same homotopy type then*

$$\mathrm{cat}_{X,Y}(X) = \mathrm{cat}_{X',Y'}(X').$$

b) If there exists a retraction $r : X' \to X$ then

$$cat_{X,Y}(A) = cat_{X',Y}(A)$$

for any closed subset A of X containing Y.

Proof: The result follows immediately from Proposition 2.6. and from the fact that $cat_{X',Y}(A) \leq cat_{X,Y}(A)$ since $X \subseteq X'$. □

Corollary 2.7.a) was obtained with A. Szulkin.

We now recall the following definition: a metric space X is an *absolute neighborhood extensor*, shortly an *ANE*, if, for every metric space E, every closed subset F of E and every map $f : F \to X$ there exists a continuous extension of f defined on a neighborhood of F in E.

Notice that, since we work in the framework of metrizable spaces, X is an *ANE* if and only if X is an absolute neighborhood retractor (see [Hu]).

It follows easily from the above definition that an *ANE* is normal. Also, it can be shown that every *ANE* is, up to homeomorphism, a retract of an open subset of a Banach space. Also, the above property is local, in the sense that X is an *ANE* if and only if X is locally an *ANE* (i.e., every point in X has a neighborhood which is an *ANE*). For those and other results, we refer the reader to [Hu]. Important examples of *ANE* are: closed convex subsets of normed spaces, Banach manifolds, manifolds with boundary, finite product of *ANE*'s and retracts of open subsets of *ANE*'s.

Lemma 2.8. *Let A be a closed subset of an ANE X. If A is contractible in X, then there exists a closed neighborhood B of A which is also contractible in X.*

Proof: Let $x_0 \in X$ and $h_t : A \to X$ be a deformation such that $h_1(x) = x_0$ for every $x \in A$. The set

$$F = ([0, 1] \times A) \cup (\{0\} \times X) \cup (\{1\} \times X)$$

is closed in $E = [0, 1] \times X$. The function $f : F \to X$ defined by

$$f(t, x) = h(t, x), \quad x \in A, t \in [0, 1]$$
$$f(0, x) = x, \quad x \in X$$
$$f(1, x) = x_0, \quad x \in X$$

is continuous. Let $\overline{f}$ be a continuous extension of f defined on a neighborhood U of F. We can assume U is closed since X is normal. Using the compactness of the interval $[0, 1]$ it is easy to verify the existence of a closed neighborhood B of A such that $[0, 1] \times B \subseteq U$, and then B is contractible in X. □

Proposition 2.9. *Let Y be a closed subset of X and suppose that both Y, X are ANE's. Then for any closed subset A of X there exists a closed neighborhood B of A such that*

$$cat_{X,Y}(A) = cat_{X,Y}(B).$$

Proof: Let $k = cat_{X,Y}(A)$ and write $A = \bigcup_{i=0}^{k}$ as in Definition 2.1 ; using the above lemma, it suffices to show that $B_0 <_Y Y$ for some closed neighborhood B_0 of A_0.

Now, since X is normal and Y is an ANE we can extend the map $h_1 : A_0 \to Y$ given by (c) of Definition 2.1 to a map $k : V \to Y$ defined on a closed neighborhood V of A_0. The set

$$F = ([0, 1] \times A) \cup (\{0\} \times V) \cup (\{1\} \times V)$$

is closed in $E = [0, 1] \times V$. Consider the map

$$f(t, x) = h(t, x), \quad x \in A, 0 \le t \le 1$$
$$f(0, x) = x, \quad x \in V$$
$$f(1, x) = k(x), \quad x \in V$$

and, as before, let $\bar{f}$ be a continuous extension of f on $[0, 1] \times B_0$, where B_0 is a closed neighborhood of A_0 contained in V. The deformation $\bar{f}_t : B_0 \to X$ immediately shows that $B_0 <_Y Y$ and this ends the proof.   □

Given two subsets A, B of a space X we call the pair (A, B) *excisive* whenever $\operatorname{int}_{A \cup B}(A) \cup \operatorname{int}_{A \cup B}(B) = A \cup B$. The following lemma will be useful in the next section:

Lemma 2.10. *Given a closed set Y of an ANE X, let $k = cat_{X,Y}(X)$. Then there exist $k+1$ subsets $A_0, \ldots, A_k$ of X such that $X = \bigcup_{i=0}^{k} A_i$, $A_0 <_Y Y$ in X, the sets $A_1, \ldots, A_k$ are contractible in X and the pairs*

$$\left(A_0, \bigcup_{i=1}^{k} A_i\right), \quad (A_i, A_j)$$

are excisive, for $i, j = 1, \ldots, k$.

Proof: Write $X = \bigcup_{i=0}^{k} B_i$ according to Definition 2.1. Choosing an open contractible neighborhood A_i of B_i, for $i = 1, \ldots, k$, and setting $A_0 = B_0$, it is immediate to check the properties stated above. Let us note that the boundary of B_0 is contained in $\bigcup_{i=1}^{k} A_i$.
□

3. Relative Cuplength

Let us recall that if X is an ANE then

$$cat_X(X) \ge 1 + \text{ cuplength } (X).$$

Using a notion of relative cuplength (introduced in [FW$_4$]) we shall obtain a similar estimate for the relative category. We use singular homology and cohomology over the real field $\mathbb{R}$ and denote it by H_* and H^* respectively.

We recall the following definition : a subset A of a topological space X is a *strong deformation retract* of X if and only if there exists a deformation $h_t : X \to X$ such that $h_1(X) \subset A$ and $h_t(x) = x$, for every $x \in A$, $t \in [0, 1]$. We state without proof the following elementary results (see e.g. [M W, ch. 8]):

Lemma 3.1. *Consider subspaces $B \subseteq A \subseteq X$.*

(a) If A is a strong deformation retract of X then

$$H_*(X, B) \simeq H_*(A, B).$$

(b) *If B is a strong deformation retract of A then*

$$H_*(X, B) \simeq H_*(X, A)$$

(c) *If A is contractible in X and i : A → X is the inclusion map then the induced morphism $i_* : H_*(A) \to H_*(X)$ is such that $i_* \equiv 0$ for $* \geq 1$.*

(d) *If A $<_B$ B in X and i : (A, B) → (X, B) is the inclusion map then the induced morphism $i_* : H_*(A, B) \to H_*(X, B)$ is such that $i_* \equiv 0$ for $* \geq 0$.*

Given a collection $B_1, \ldots, B_n$ of subspaces of a space X such that the pairs (B_i, B_j) are excisive we denote by $\cup$ the cup product either as a n-multilinear map $\oplus_{i=1}^n H^*(X, B_i) \to H^*(X, \bigcup_{i=1}^n B_i)$ or as a linear map $\otimes_{i=1}^n H^*(X, B_i) \to H^*(X, \bigcup_{i=1}^n B_i)$.

Definition 3.2. *Let Y be a closed subset of a space X. The cuplength of X relative to Y is the largest integer n such that there exist $\alpha_0 \in H^*(X, Y)$, $* \geq 0$ and $\alpha_1, \ldots, \alpha_n \in H^*(X)$, $* \geq 1$ with*

$$\alpha_0 \cup \alpha_1 \cup \ldots \cup \alpha_n \neq 0.$$

We write then cuplength $(X, Y) = n$. We set cuplength $(X, Y) = -\infty$ if no such integer exists.

Remark 3.3.

(i) It follows from the definition that cuplength $(X, Y) \geq 0$ if and only if $H^k(X, Y)$ is nontrivial for some $k \geq 0$.

(ii) Since the cup product in $\oplus_{n \geq 0} H^n(X)$ has a 0-cochain as identity element we see that cuplength $(X, \phi) =$ cuplength (X).

Example 3.4. Let $B = \{x \in \mathbb{R}^N : \| x \| \leq 1\}$. Then

$$\text{cuplength } (B, \partial B) = 0 \quad , \quad \text{for } N \geq 1;$$
$$\text{cuplength } (\partial B) = 1 \quad , \quad \text{for } N \geq 2.$$

This follows readily from the simple well-known facts:

$$H^*(B) \simeq \mathbb{R} \quad \text{for} \quad * = 0;$$
$$H^*(\partial B) \simeq \mathbb{R} \quad \text{for} \quad * = 0, N \ (N \geq 2);$$
$$H^*(B, \partial B) \simeq \mathbb{R} \quad \text{for} \quad * = N,$$

while they are trivial otherwise.

We shall see that the relative cuplength is less precise but easier to compute than the relative category.

Proposition 3.5. *Let X, V be topological spaces and Y be a closed subspace of X. If $H^*(X, Y)$ or $H^*(V)$ is of finite type then*

$$\text{cuplength } (X \times V, Y \times V) = \text{cuplength } (X, Y) + \text{cuplength } (V).$$

Proof: We introduce projections

$$
\begin{aligned}
p_1 &: \quad X \times V \to X, \\
p_2 &: \quad X \times V \to V, \\
p_3 &: \quad (X, Y) \times V \to (X, Y).
\end{aligned}
$$

Let $n =$ cuplength (X, Y) and $k =$ cuplength (V). We recall the Künneth formula : under the above assumptions there exists an isomorphism

$$\left(H^*(X, Y) \otimes H^*(V)\right)^q \xrightarrow{\;\;K\;\;} H^q(X \times V, Y \times V)$$
$$\simeq H^q((X, Y) \times V)$$

given by

$$K(a \otimes b) = p_3^*(a) \cup p_2^*(b).$$

Also, the following diagram is commutative:

$$
\begin{array}{ccc}
H^p(X, Y) \oplus H^q(X) & \xrightarrow{\;\;\cup\;\;} & H^{p+q}(X, Y) \\
{\scriptstyle (p_3^*, p_1^*)} \downarrow & & \downarrow {\scriptstyle p_3^*} \\
H^p(X \times V, Y \times V) \oplus H^q(X \times V) & \xrightarrow{\;\;\cup\;\;} & H^{p+q}(X \times V, Y \times V)
\end{array}
$$

(see [Sp], [Do]).

By definition, there exists $\alpha_0 \in H^*(X, Y)$, $* \geq 0$ and $\alpha_1, \cdots, \alpha_n \in H^*(X)$, $* \geq 1$, such that

$$\alpha_0 \cup \alpha_1 \cup \cdots \cup \alpha_n \neq 0.$$

There exists also $\beta_1, \cdots, \beta_k \in H^*(V)$, $* \geq 1$, such that

$$\beta_1 \cup \cdots \cup \beta_k \neq 0.$$

Then we obtain

$$
\begin{aligned}
0 \neq &\; K[(\alpha_0 \cup \ldots \cup \alpha_n) \otimes (\beta_1 \cup \ldots \cup \beta_k)] \\
= &\; p_3^*(\alpha_0 \cup \ldots \cup \alpha_n) \cup p_2^*(\beta_1 \cup \ldots \cup \beta_k) \\
= &\; p_3^*(\alpha_0) \cup p_1^*(\alpha_1 \cup \ldots \cup \alpha_n) \cup p_2^*(\beta_1 \cup \ldots \cup \beta_k) \\
= &\; p_3^*(\alpha_0) \cup p_1^*(\alpha_1) \cup \ldots \cup p_1^*(\alpha_n) \cup p_2^*(\beta_1) \cup \ldots \cup p_2^*(\beta_k).
\end{aligned}
$$

From the definition of relative cuplength we then get

$$\text{cuplength } (X \times V, Y \times V) \geq n + k.$$

The reversed inequality follows directly from the Künneth formula.

The relation between relative cuplength and relative category is given by the following

Theorem 3.6. *Let Y be a closed subset of an ANE X. Then*

$$cat_{X,Y}(X) \geq 1 + \text{cuplength } (X, Y).$$

Proof: Let $k = cat_{X,Y}(X)$; we may assume k is finite. Write $X = \bigcup_{i=0}^{k} A_i$ according to lemma 2.10.

Consider the inclusions

$$i_0 : (A_0, Y) \to (X, Y), \quad \ell_0 : (X, Y) \to (X, A_0)$$
$$i_j : A_j \to X, \quad \ell_j : (X, \phi) \to (X, A_j)$$

$$\theta_1 : (X, Y) \to (X, X), \quad \theta_2 : (X, \phi) \to (X, \bigcup_{i=1}^{k} A_i).$$

Then the sequences :

$$\ldots \to H^*(X, A_j) \xrightarrow{\ell_j^*} H^*(X) \xrightarrow{i_j^*} H^*(A_j) \to \ldots$$

$$\ldots \to H^*(X, A_0) \xrightarrow{\ell_0^*} H^*(X, Y) \xrightarrow{i_0^*} H^*(A_0, Y) \to \ldots$$

are exact and the following diagrams are commutative

$$
\begin{array}{ccc}
H^*(X, A_0) \oplus H^*(X, \bigcup_{i=1}^{k} A_i) & \xrightarrow{\cup} & H^*(X, X) \simeq \{0\} \\
\downarrow{\scriptstyle \ell_0^* \oplus \theta_2^*} & & \downarrow{\scriptstyle \theta_1^*} \\
H^*(X, Y) \oplus H^*(X) & \xrightarrow{\cup} & H^*(X, Y)
\end{array}
$$

$$
\begin{array}{ccc}
H^*(X, A_1) \oplus \ldots \oplus H^*(X, A_k) & \xrightarrow{\cup} & H^*(X, \bigcup_{i=1}^{k} A_i) \\
\downarrow{\scriptstyle L^*} & & \downarrow{\scriptstyle \theta_2^*} \\
H^*(X) \oplus \ldots \oplus H^*(X) & \xrightarrow{\cup} & H^*(X)
\end{array}
$$

where $L = \ell_1^* \oplus \ldots \oplus \ell_k^*$. We deduce from lemma 3.1 that $i_0^* = 0$ for $* \geq 0$ and $i_j^* = 0$ for $* \geq 1, j = 1, \ldots, k$; by exactness ℓ_0^* is surjective for $* \geq 0$ and L^* is surjective for $* \geq 1$.

Now, given $\beta_0, \ldots, \beta_k$ as in Definition 3.2, let $\alpha_0 \in H^*(X, A_0)$, $\alpha_i \in H^*(X, A_i)$ be such that $\ell_0^*(\alpha_0) = \beta_0$, $\ell_i^*(\alpha_i) = \beta_i$; we write $\beta := \beta_1 \cup \ldots \cup \beta_k \in H^*(X)$, $\alpha := \alpha_1 \cup \ldots \cup \alpha_k \in H^*(X, \bigcup_{i=1}^{k} A_i)$. Then

$$
\begin{aligned}
\beta_0 \cup \beta_1 \cup \ldots \cup \beta_k &= \beta_0 \cup \beta \\
&= \ell_0^*(\alpha_0) \cup \theta_2^*(\alpha) \\
&= \theta_1^*(\alpha_0 \cup \alpha) \\
&= \theta_1^*(0) \\
&= 0.
\end{aligned}
$$

Since the sequence $\beta_0, \ldots, \beta_k$ was arbitrary we deduce

$$\text{cuplength } (X, Y) \leq k - 1$$

and this proves the theorem.

Example 3.7.

a) Let V be an ANE and $B := \{x \in \mathbb{R}^N : \| x \| \leq 1\}$, $N \geq 1$. Then

$$\text{cat}_{B \times V, \partial B \times V}(B \times V) \geq 1 + \text{cuplength } (V).$$

b) Let $A := \{x \in \mathbb{R}^N : 1 \leq \| x \| \leq 2\}$, $N \geq 1$. Then

$$\text{cat}_{A, \partial A}(A) = 2.$$

Indeed, assume $N \geq 2$ (the other case is simpler). The homeomorphism $h : A \to$ $[0, 1] \times S^{N-1}$, $h(x) = \left(\| x \| - 1, \frac{x}{\|x\|} \right)$ shows that

$$
\begin{aligned}
\text{cat}_{A, \partial A}(A) &= \text{cat}_{[0,1] \times S^{N-1}, \{0,1\} \times S^{N-1}}([0, 1] \times S^{N-1}) \\
&\geq 1 + \text{cuplength } (S^{N-1}) + \text{cuplength } ([0, 1], \{0, 1\}) \\
&= 2.
\end{aligned}
$$

and the conclusion follows from Example 2.2.(a).

4. Limit Relative Category

In this section we consider a topological space X together with a sequence $(X_n)_{n \geq 1}$ of closed subsets of X. We assume that there exists, for every n, a retraction $r_n : X \to X_n$. If A is any subspace of X we denote by A_n the set $A \cap X_n$.

Definition 4.1. Let $Y \subseteq A$ be closed subsets of X. The limit relative category of A in X relative to Y, with respect to (X_n), is defined by

$$\text{cat}^\infty_{X,Y}(A) = \limsup_{n \to \infty} \text{cat}_{X_n, Y_n}(A_n).$$

We point out that it is not true in general that $\text{cat}^\infty_{X,Y}(A) \leq \text{cat}_{X,Y}(A)$; but this will be the case if $r_n(Y) \subseteq Y$ for every n large.

If A, B, Y are closed subsets of Y then, by definition, $A <^\infty_Y B$, with respect to (X_n), if and only if $Y \subseteq A \cap B$ and $A_n <_{Y_n} B_n$ for every n large.

Proposition 4.2. Let A, B, Y be closed subsets of X with $Y \subseteq A \cap B$. The limit relative category satisfies the following properties:

a) Normalization : $\text{cat}^\infty_{X,Y}(Y) = 0$;
b) Subadditivity : $\text{cat}^\infty_{X,Y}(A \cup B) \leq \text{cat}^\infty_{X,Y}(A) + \text{cat}_X(B)$;
c) Homotopy : if $A <^\infty_Y B$ then $\text{cat}^\infty_{X,Y}(A) \leq \text{cat}^\infty_{X,Y}(B)$;
d) Monotonicity : if $A \subseteq B$ then $\text{cat}^\infty_{X,Y}(A) \leq \text{cat}^\infty_{X,Y}(B)$.

Proof: This follows readily from Proposition 2.4. As for b), note that a subset of X_n is contractible in X_n if and only if it is contractible in X ; hence

$$\text{cat}_{X_n}(B_n) = \text{cat}_X(B_n) \leq \text{cat}_X(B).$$

Example 4.3. Let E be a normed space such that $E = W \oplus Z$ (topological direct sum) and for some $0 < r_1 < r_2$, define $A = \{w \in W : r_1 \leq \| w \| \leq r_2\}$, $\partial A = \{w \in W : \| w \| = r_1$

or $\| w \| = r_2 \}$. Then, with respect to any sequence of vector subspaces $E_n = W_n \oplus Z_n$, $W_n \subseteq W$, $Z_n \subseteq Z$, $1 \le \dim W_n < \infty$ we have

$$\mathrm{cat}^{\infty}_{A \times Z, \partial A} A = 2.$$

This follows from corollary 2.7.b) and Example 3.7.b).

Example 4.4. Let E, E_n be as above, V be an ANE and $X := E \times V$. Define $B = \{w \in W : \| w \| \le R \}$, $\partial B = \{w \in W : \| w \| = R \}$. Then, with respect to $X_n := E_n \times V$

$$\mathrm{cat}^{\infty}_{X, \partial B \times V} (B \times V) \ge 1 + \text{ cuplength } (V).$$

This follows from Corollary 2.7.b) and Example 3.7.a).

5.   The Deformation Lemma

In the sequel we let X be a connected Finsler manifold of class C^1. We refer the reader to [Pa] and [Pa$_1$] for the definition of Finsler structure and some of its properties. For the applications we have in mind X could be a connected Riemannian manifold of class C^1 (see [MW]). For the reader's convenience we collect here some basic and classical results on the method of "descent flow" that are on the basis of our minimax principle using relative category.

We denote the tangent bundle of X by $T(X)$ and the tangent space of X at a point $x \in X$ by $T_x(X)$. Recall that the cotangent bundle $T(X)^*$ has a dual Finsler structure given by

$$\| w \| = \sup \{ \langle w, v \rangle : v \in T_x(X), \| v \| = 1 \}$$

where $w \in T_x(X)^*$ and $\langle \cdot, \cdot \rangle$ is the duality pairing between $T_x(X)^*$ and $T_x(X)$.

Let $f \in C^1(X; \mathbb{R})$ and denote the differential of f at x by $df(x)$; then $df(x)$ is an element of $T_x(X)^*$. A point $x \in X$ is said to be a *critical point* of f if $df(x) = 0$. The corresponding value $c = f(x)$ will be called a *critical value* for f. We shall repeatedly use the following notation:

$$K = \{x \in X : df(x) = 0\} \quad , \quad K_c = K \cap f^{-1}(c)$$
$$f^c = \{x \in X : f(x) \le c\} \quad , \quad S_\delta = \{x \in X : d(x, S) \le \delta\}$$

where $\delta > 0$ and $S \subseteq X$ is any subset (recall that X is a metric space, cf. [Pa$_1$]).

A map $V : X \backslash K \to TX$ is said to be a pseudo gradient vector field for f if V is locally Lipschitz continuous and

$$\| V(x) \| \le 2 \| df(x) \|, \langle df(x), V(x) \rangle \ge \| df(x) \|^2$$

for every $x \in X \backslash K$. It can be proved that every C^1 function admits a pseudo-gradient vector field, if X is a C^2 Finsler manifold (cf. [Pa]).

Lemma 5.1. *Let g be a locally Lipschitz continuous vector field on a C^2 Finsler manifold X. Then, for any $x \in X$, the Cauchy problem*

$$\begin{aligned} \sigma(t) &= g(\sigma(t)) \\ \sigma(0) &= x \end{aligned}$$

has a unique maximal solution $\sigma(\cdot, x)$ *defined on some open interval* $]w_-(x), w_+(x)[$
containing 0. *The set* $\mathcal{D} = \{(t, x) : w_-(x) < t < w_+(x)\}$ *is open in* $\mathbb{R} \times X$ *and the flow*
$\sigma : \mathcal{D} \to X,\ (t, x) \to \sigma(t, x)$ *is continuous. Moreover, if* $\{\sigma(t, x) : w_-(x) < t < w_+(x)\}$
is contained in a complete subset of X *on which the function* g *is bounded then* $w_-(x) =$
$-\infty,\ w_+(x) = +\infty.$

For a proof see [Pa]. The following result is a quantitative version due to Willem [Wi]
of Clark's deformation lemma (see [Ra]). We only sketch the proof.

Lemma 5.2. *Let* f *be a* C^1 *function defined on a* C^2 *Finsler manifold* X *and let* $c \in \mathbb{R}$,
$\varepsilon > 0$ *be such that* $f^{-1}([c - 2\varepsilon, c + 2\varepsilon])$ *is complete. Then, given any set* $S \subseteq X$ *and
any* $\delta > 0$ *such that* $\| df(x) \| \geq \frac{4\varepsilon}{\delta}$ *for every* $x \in S_{2\delta} \cap f^{-1}([c - 2\varepsilon, c + 2\varepsilon])$, *there exists
a deformation* $h_t : X \to X$ *such that*

- $f \circ h(\cdot, x)$ *is non-increasing,* $\forall x \in X$;
- $h_1(f^{c+\varepsilon} \cap S) \subseteq f^{c-\varepsilon} \cap S_\delta$;
- $h_t(x) = x$ *for every* $x \in X \backslash (S_{2\delta} \cap f^{-1}([c - 2\varepsilon, c + 2\varepsilon]))$.

Proof: Denote

$$A := S_{2\delta} \cap f^{-1}([c - 2\varepsilon, c + 2\varepsilon]),$$
$$B := f^{-1}([c - \varepsilon, c + \varepsilon])$$

and let $\psi : X \to [0, 1]$ be the locally Lipschitz continuous function given by

$$\psi(x) = \frac{d(x, X \backslash A)}{d(x, X \backslash A) + d(x, B)}.$$

Choose a pseudo-gradient vector field $V : K \to T(X)$ for f and define $g : X \to T(x)$ by

$$g(x) = -\psi(x) \frac{V(x)}{\| V(x) \|}$$

for $x \in A$, and $g(x) := 0$ for $x \in X \backslash A$. Then g is a bounded locally Lipschitz continuous
vector field on X.

By Lemma 5.1 the corresponding Cauchy problem has, for any $x \in X$, an unique
solution $\sigma(\cdot, x)$ defined on $\mathbb{R}$. Letting

$$h(t, x) := \sigma(\delta t, x)$$

it is easy to check that h_t has the desired properties.

Consider now a given sequence (X_n) of closed, connected submanifolds of class C^2
of the C^1 manifold X. Denoting $f_n := f |_{X_n}$ we then have $f_n \in C^1(X_n; \mathbb{R})$, $n \geq 1$.

Definition 5.3. *Given* $c \in \mathbb{R}$ *we say that* f *satisfies the Palais-Smale condition with
respect to* (X_n) *at level* c *if every sequence* (x_{nj}) *satisfying*

$$n_j \to \infty\ and\ x_{nj} \in X_{nj},\ f(x_{nj}) \to c,\ \| df_{nj}(x_{nj}) \| \to 0,$$

possesses a subsequence which converges in X *to a critical point of* f. *The above
property will be referred as the* $(PS)^*_c$ *condition with respect to* (X_n).

Proposition 5.4. *Let $c \in \mathbb{R}$, $f \in C^1(X, \mathbb{R})$ and $Y \subset X$ be closed. Assume that*

a) $\sup_Y f < c$;
b) f satisfies the $(PS)_c^*$ condition with respect to (X_n);
c) there exists $\eta > 0$ such that $f^{-1}([c - \eta, c + \eta])$ is complete.

Then, for every open neighborhood N of K_c such that $N \cap Y = \phi$, there exists $\varepsilon > 0$ such that

$$f^{c+\varepsilon} \backslash N <_Y^\infty f^{c-\varepsilon}.$$

Proof: Let N be an open neighborhood of K_c such that $N \cap Y = \phi$. The $(PS)_c^*$ condition implies the existence of $\delta > 0$ and $n_0 \in \mathbb{N}$ such that $\| df_n(x) \| \geq \delta$, for every $n \geq n_0$, and

$$x \in f_n^{-1}([c - 2\delta, c + 2\delta]) \cap (X_n \backslash N)_{2\delta}.$$

Choose $\varepsilon = \min\{\eta/2, \delta/4, \delta, (c - \sup_Y f)/4\}$ and define, for $n \geq n_0$ fixed, $S = X_n \backslash N$. According to lemma 5.2, there exists a deformation $h_t : X_n \to X_n$ such that

$$h_1(f_n^{c+\varepsilon} \cap S) \subseteq f_n^{c-\varepsilon} \cap S_\delta,$$

$$h_t(x) = x \text{ for every } x \in X_n \backslash (S_{2\delta} \cap f_n^{-1}([c - 2\varepsilon, c + 2\varepsilon])).$$

In particular, we have

$$h_1(f_n^{c+\varepsilon} \backslash N) \subseteq f_n^{c-\varepsilon},$$

$$h_t(y) = y \text{ for every } y \in Y_n,$$

so that, by definition, $f^{c+\varepsilon} \backslash N <_Y^\infty f^{c-\varepsilon}$.

6. Critical Point Theorems

In this section, we consider a connected Finsler manifold X of class C^1 together with a sequence $(X_n)_{n \geq 0}$ of closed, connected submanifolds of class C^2 of X. We assume that there exists, for every $n \geq 0$, a retraction $r_n : X \to X_n$; the limit relative category is computed with respect to (X_n).

Let $f \in C^1(X; \mathbb{R})$ and Y be a closed subset of X. Define, for each $j \geq 1$,

$$\mathcal{A}_j = \{A \subseteq X : A \text{ is closed }, A \supseteq Y, \text{cat}_{X,Y}^\infty(A) \geq j\},$$
$$c_j = \inf_{A \in \mathcal{A}_j} \sup_{x \in A} f(x).$$

Theorem 6.1. *Assume that*

a) $\sup_Y f < c_k = c_{k+1} = \ldots = c_{k+m} =: c < +\infty$;
b) f satisfies the $(PS)_c^*$ condition with respect to (X_n);
c) there exists $\eta > 0$ such that $f^{-1}([c - \eta, c + \eta])$ is complete;

then c is a critical value of f and

$$\text{cat}_X(K_c) \geq m + 1.$$

Proof: Proposition 2.9. implies the existence of an open neighborhood N of K_c such that $\mathrm{cat}_X(\overline{N}) = \mathrm{cat}_X(K_c)$. Using assumption a), we can assume that $N \cap Y = \phi$. From proposition 5.4. and assumptions a), b) c), there exists $\varepsilon > 0$ such that

$$f^{c+\varepsilon} \backslash N <_Y^\infty f^{c-\varepsilon}.$$

Using Proposition 4.2. and the definitions of c_{k+m} and c_k, we obtain

$$
\begin{aligned}
k + m &\leq \mathrm{cat}_{X,Y}^\infty(f^{c+\varepsilon}) \\
&\leq \mathrm{cat}_{X,Y}^\infty(f^{c+\varepsilon} \backslash N) + \mathrm{cat}_X(\overline{N}) \\
&\leq \mathrm{cat}_{X,Y}^\infty(f^{c-\varepsilon}) + \mathrm{cat}_X(K_c) \\
&\leq k - 1 + \mathrm{cat}_X(K_c).
\end{aligned}
$$

Remark 6.2.

1. The following example satisfies a), b) and $c_1 = 0$ is not a critical value:

$$X =]0, +\infty[, \ Y = \phi, \ f(x) = x.$$

2. The following example satisfies a), c) and $c_1 = 0$ is not a critical value:

$$X = \mathbb{R}, \ Y = \phi, \ f(x) = e^x.$$

We consider now the generalized Saddle Point Theorem. We assume that $X = E \times V$ where E is a Banach space and V is a complete connected Finsler manifold of class C^2. Let $E = W \oplus Z$ (topological direct sum) and $E_n = W_n \oplus Z_n$ be a sequence of closed subspaces with $Z_n \subseteq Z$, $W_n \subseteq W$, $1 \leq \dim W_n < \infty$. Define

$$X_n := E_n \times V.$$

Theorem 6.3. *Assume there exist $r > 0$ and $\alpha < \beta \leq \gamma$ such that*

a) f satisfies the $(PS)_c^*$ condition with respect to (X_n) for every $c \in [\beta, \gamma]$;
b) $f(w, v) \leq \alpha$ for every $(w, v) \in W \times V$ such that $\| w \| = r$;
c) $f(z, v) \geq \beta$ for every $(z, v) \in Z \times V$;
d) $f(w, v) \leq \gamma$ for every $(w, v) \in W \times V$ such that $\| w \| \leq r$.
Then $f^{-1}([\beta, \gamma])$ contains at least cuplength $(V) + 1$ critical points of f.

Proof: We apply Theorem 6.1 with

$$Y := \{w \in W : \| w \| = r\} \times V.$$

Let us define $m := \mathrm{cuplength} \, (V)$ and

$$A := \{w \in W : \| w \| \leq r\} \times V.$$

It follows from Example 4.4. that

$$\mathrm{cat}_{X,Y}^\infty(A) \geq m + 1.$$

Thus $A \in \mathcal{A}_j$, $j = 1, \ldots, m + 1$. Assumption (d) implies

$$c_1 \leq c_2 \leq \ldots \leq c_{m+1} \leq \gamma.$$

Assume now that $\sup_B f < \beta$ for some $B \in \mathcal{A}_1$. Assumption (c) implies that $B \cap (Z \times V) = \phi$. Thus the deformations $h_n : [0, 1] \times B_n \to X_n$ given by

$$h_n(t, w, z, v) = \left(((1-t) + \frac{tr}{\| w \|})w, (1-t)z, v \right)$$

are well-defined and show that $\mathrm{cat}_{X,Y}^\infty(B) = 0$, contradicting the definition of $\mathcal{A}_1$. We conclude from assumption b) that

$$\sup_Y f \le \alpha < \beta \le c_1$$

and the theorem follows.   □

Remark 6.4. When V is a singleton and W is finite dimensional we obtain the Saddle Point Theorem of P. Rabinowitz [Ra]. Theorem 6.3. was proved by J.Q. Liu ([Li], Th. 3.2.) under the further assumption that f and $f \mid_{E_n}$ satisfy the Palais-Smale condition on $E \times V$ and $E_n \times V$ respectively. A similar theorem was recently obtained by A. Szulkin ([Sz, Th. 3.8.]), replacing the $(PS)^*$ condition by some suitable compactness assumption on the functional f.

We consider now the generalized linking theorem. Let X, X_n be as above. Let $R > 0$, $r > 0$, $\rho \in]0, r[$ and suppose $e \in \bigcap_{n=0}^\infty Z_n$, $\| e \| = 1$. Define

$$Q := \{w \in W : \| w \| \le R\} \oplus \{\lambda e : 0 \le \lambda \le r\},$$
$$\partial Q := \{w \in W : \| w \| = R\} \oplus \{\lambda e : 0 \le \lambda \le r\}$$
$$\cup \{w \in W : \| w \| \le R\} \oplus \{0, re\}$$

Theorem 6.5. *Assume there exist $\alpha < \beta \le \gamma$ such that*

a) f satisfies the $(PS)_c^*$ condition with respect to (X_n) for every $c \in [\beta, \gamma]$;
b) $f(x, v) \le \alpha$ for every $(x, v) \in \partial Q \times V$;
c) $f(z, v) \ge \beta$ for every $(z, v) \in Z \times V$ such that $\| z \| = \rho$;
d) $f(x, v) \le \gamma$ for every $(x, v) \in Q \times V$.
Then $f^{-1}([\beta, \gamma])$ contains at least cuplength $(V) + 1$ critical points of f.

Proof: We apply theorem 6.1. with $Y := \partial Q \times V$. It is easy to see that $\mathrm{cat}_{X,Y}^\infty (Q \times V) \ge$ cuplength $(V) + 1$. Assume that $\sup_{\mathcal{B}} f < \beta$ for some $B \in \mathcal{A}_1$. Assumption c) implies that $B \cap (\{z \in Z : \| z \| = \rho\} \times V) = \phi$. Let $\theta_n : W_n \oplus \{\lambda e : \lambda \in \mathbb{R}\} \setminus \{\rho e\} \to \partial Q \cap E_n$ be a retraction. Then the deformations $h_n : [0, 1] \times B_n \to X_n$ given by

$$h_n(t, w, z, v) = ((1-t)(w+z) + t\theta_n(w+ \| z \| e), v)$$

are well-defined and show that $\mathrm{cat}_{X,Y}^\infty(B) = 0$, contradicting the definition of $\mathcal{A}_1$. The argument then follows the one of the preceeding proof.   □

Remark 6.6. When V is a singleton and W is finite dimensional we obtain the linking theorem in [Ra, Th. 5.3.]. It contains as a particular case the Mountain-Pass Theorem of Ambrosetti-Rabinowitz, setting $W = \{0\}$; for other related connectedness theorems of the type, see [FW$_2$].

Our final result concerns the Mountain Circle Theorem first proved in [FW$_3$] when W is finite dimensional. Let E, E_n be as above.

Theorem 6.7. *Assume there exist constants* $0 < r_1 < r_2 < r_3$, $\alpha < \beta \le \gamma$ *such that*

a) *f satisfies the $(PS)_c^*$ condition with respect to (E_n) for every $c \in [\beta, \gamma]$;*
b) *$f(w, z) \le \alpha$ for every $(w, z) \in W \times Z$, $\| w \| \le r_1$;*
c) *$f(w, z) \ge \beta$ for every $(w, z) \in W \times Z$, $\| w \| = r_2$;*
d) *$f(w, 0) \le \alpha$ for every $w \in W$, $\| w \| = r_3$;*
e) *$f(w, 0) \le \gamma$ for every $w \in W$, $\| w \| \le r_3$.*

Then $f^{-1}([\beta, \gamma])$ contains at least two critical points of f.

Proof: Define $X := \{w \in W : \| w \| > r_1/2\} \times Z$, $X_n := X \cap E_n$ and $\tilde{f} := f \mid_X$. We apply theorem 6.1. with $Y := \{w \in W : \| w \| = r_1 \text{ or } \| w \| = r_3\}$. Define $A := \{w \in W : r_1 \le \| w \| \le r_3\}$. It follows from example 4.3. that

$$\text{cat}_{X,Y}^\infty(A) = \text{cat}_{A \times Z, Y}^\infty(A) = 2.$$

Assumption b) implies that $\tilde{f}^{-1}\{[(\alpha + \beta)/2, +\infty[\}$ is complete. It follows from assumption e) that $c_1 \le c_2 \le \gamma$. Assume that $\sup_B \tilde{f} < \beta$ for some $B \in \mathcal{A}_1$. Assumption c) implies that $\| w \| \ne r_2$ for every $(w, z) \in B$. The deformations $h_n : [0, 1] \times B_n \to X_n$ given by

$$
\begin{aligned}
h_n(t, w, z) &= ((1 - t) + t r_3/ \| w \|)w + (1 - t)z \text{ if } \| w \| > r_2 \\
&= ((1 - t) + t r_1/ \| w \|)w + (1 - t)z \text{ if } \| w \| < r_2
\end{aligned}
$$

are well-defined and show that $\text{cat}_{X,Y}^\infty(B) = 0$, contradicting the definition of $\mathcal{A}_1$. We conclude from assumptions b) and d) that

$$\sup_Y \tilde{f} \le \alpha < \beta \le c_1.$$

$\square$

7. Some Applications

As a first application we consider the periodic problem

$$
\begin{aligned}
Jz + \nabla H(t, z) &= h(t) \\
z(0) &= z(T)
\end{aligned}
\tag{HS}
$$

where $T > 0$, $z = (p, q) \in \mathbb{R}^N \times \mathbb{R}^N$, $J = \begin{bmatrix} 0 & -I \\ I & 0 \end{bmatrix}$ is the standard sympletic matrix (I denotes the identity matrix in $\mathbb{R}^N$), $h \in \mathcal{C}([0, T]; \mathbb{R}^{2N})$, $H \in \mathcal{C}([0, T] \times \mathbb{R}^{2N}; \mathbb{R})$ and has continuous partial derivatives with respect to z; $\nabla H(\cdot, \cdot)$ denotes the gradient with respect to z.

We assume there exists a symmetric matrix $A_\infty(t)$ which is continuous in t and such that

$$\nabla H(t, z) = A_\infty(t)z + o(| z |) \tag{H1}$$

as $| z | \to \infty$, uniformly in t.

We denote by K the finite dimensional space of solutions of the periodic problem

$$Jz + A_\infty(t)z = 0$$
$$z(0) = z(T)$$

and suppose

$$K = \text{span } \{v_1, \ldots, v_k\} \subseteq \mathbb{R}^{2N} \qquad (H2)$$

Let us denote by G the additive group

$$G := \{\sum_{j=1}^{k} b_j v_j : b_j \in \mathbb{Z}, 1 \leq j \leq k\}.$$

Two solutions z_1, z_2 of (HS) are geometrically distinct if $z_1 - z_2 \notin G$.

Theorem 7.1. *Assume (H1), (H2) hold and moreover*

$$H(\cdot, z + v_j) = H(\cdot, z), \quad \int_0^T (h(t), v_j) \, dt = 0. \qquad (H3)$$

for any $z \in \mathbb{R}^{2N}$, $j = 1, \ldots, k$. Then problem (HS) has at least $k + 1$ geometrically distinct solutions.

Proof:

1) Let us describe the functional framework. After a change of variable, we can assume that $T = 1$.
Let $L^2(\mathbb{T}, \mathbb{R}^{2N})$ be the space of square integrable functions defined on $\mathbb{T} = \mathbb{R}/\mathbb{Z}$ with values in $\mathbb{R}^{2N}$.
Each function $u \in L^2(\mathbb{T}, \mathbb{R}^{2N})$ has a Fourier expansion

$$\sum_{k \in \mathbb{Z}} \hat{u}(k) e^{2i\pi kt}.$$

For $s \geq 0$, let us define the space

$$H^s(\mathbb{T}, \mathbb{R}^{2N}) = \{u \in L^2(\mathbb{T}, \mathbb{R}^{2N}) : \sum_{k \in \mathbb{Z}} (1 + k^2)^s (\hat{u}(k))^2 < \infty\}.$$

The space $H^s(\mathbb{T}, \mathbb{R}^{2N})$ with the norm

$$\|u\|_s = \left(\sum_{k \in \mathbb{Z}} (1 + k^2)^s |\hat{u}(k)|^2 \right)^{1/2}$$

is a Hilbert space. By the usual Rellich theorem, the space $H^{1/2} = H^{1/2}(\mathbb{T}, \mathbb{R}^{2N})$ is compactly embedded in $L^p(\mathbb{T}, \mathbb{R}^{2N})$ for $p \in [1, \infty[$. We define on $H^{1/2}$ the bilinear form

$$a(u, v) = \langle u, Jv \rangle - \int_{\mathbb{T}} (A_\infty(t)u, v) \, dt,$$

where $\langle \cdot, \cdot \rangle$ is the duality pairing between $H^{-1/2}(\mathbb{T}, \mathbb{R}^{2N})$ and $H^{1/2}(\mathbb{T}, \mathbb{R}^{2N})$. We also define on $H^{1/2}$ the functional

$$\psi(u) = \int_{\mathbb{T}} [H(t, u) - (A_\infty(t)u, u) - (h(t), u)]dt.$$

Assumption (H1) implies that

$$\psi'(u) = o(|u|), \ |u| \to \infty, \ u \in H^{1/2}. \tag{1}$$

It is easy to verify (see e.g. [R_a, Appendix B]) that the solutions of (HS) are precisely the critical points of $\frac{1}{2}a(u, u) - \psi(u)$.

2) From (H2) and (H3) we deduce

$$a(u + v_j, u + v_j) = a(u, u),$$

$$\psi(u + v_j) = \psi(u),$$

for $j = 1, \cdots, k$. Thus the natural space associated to (H5) is given by

$$H^{1/2}/G \simeq E \times \mathbb{T}^k$$

where E is the orthogonal complement of K in $H^{1/2}$ and $\mathbb{T}^k$ is the k-torus. We shall apply theorem 6.3 to the functional

$$f(u, v) = \frac{1}{2}a(u, u) - \psi(u + v)$$

defined on the space $X = E \times \mathbb{T}^k$.

3) Since, by (H2), the quadratic form a is non-degenerate on E there exists an orthogonal decomposition $W \oplus Z$ of E such that, for some $\mathcal{V} > 0$,

$$a(u, u) \geq \mathcal{V}\|u\|^2, \forall u \in Z,$$

$$a(u, u) \leq -\mathcal{V}\|u\|^2, \forall u \in W$$

Let $(e_n)_{n \geq 1}$ be an hilbertean basis for W and set

$$E_n = sp(e_1, \cdots, e_n) \oplus Z,$$

$$X_n = E_n \times \mathbb{T}^k.$$

4) We claim that f satisfies the $(PS)^*_c$ with respect to (X_u) for every $c \in \mathbb{R}$. Consider a sequence $((u_{n_j}, v_{n_j}))$ such that $n_j \to \infty$ and

$$u_{n_j} \in E_{n_j}, \ \|f'_{n_j}(u_{n_j}, v_{n_j})\| \to 0. \tag{2}$$

For n large enough (1) and (2) imply, with $u_{n_j} = w + z$ and $v_{n_j} = v$

$$a(z, z) = \langle f'(w + z, v), z \rangle + \langle \psi'(w + z + v), z \rangle$$

$$\leq \|z\| + \frac{\mathcal{V}}{2}\|w + z + v\|\|z\|$$

and, similarly,

$$-a(w, w) \le \|w\| + + \frac{\nu}{2}\|w + z + v\|\|w\|,$$

so that (u_{n_j}) is bounded in E. Going if necessary to a subsequence, we can assume that $u_{n_j} \rightharpoonup u$ in E and $v_{n_j} \to v$ in $\mathbb{T}^k$. Notice that

$$a(z_{n_j} - z, z_{n_j} - z)$$

$$= \langle f'(u_{n_j}, v_{n_j}) - f'(u, v), z_{n_j} - z \rangle$$

$$+ \langle \psi'(u_{n_j} + v_{n_j}) - \psi'(u + v), z_{n_j} - z \rangle$$

where z_{n_j} (resp. z) denote the projection of u_{n_j} (resp. u) on Z. Using the fact that $u_{n_j} \to u$ in L^2, it is easy to verify that $z_{n_j} \to z$ in E. Similarly, using obvious notations, $w_{n_j} \to w$ in E. It follows then that $u_{n_j} \to u$ in X and $f'(u, v) = 0$.

5) It follows from (1) that

$$\psi(u + v) = o(|u|^2), \ |u| \to \infty, u \in E, v \in \mathbb{T}^k.$$

Hence f is bounded below on $Z \times \mathbb{T}^k$ and

$$f(w, v) \to -\infty, |w| \to \infty, w \in W, v \in \mathbb{T}^k.$$

It is then easy to verify assumptions b) and c) of theorem 6.3. Assumption d) is also satisfied since f maps bounded sets into bounded sets.

Finally, proposition 3.5 and exemple 3.4 imply

$$\text{cuplength } (\mathbb{T}^k) = k$$

and the result follows from theorem 6.3.

$$\square$$

Remark 7.2.

(a) The above theorem generalizes Theorem 5 of [Ch] and Theorem 4 of [FM$_4$] where it is assumed the existence of a bounded second derivative for H. It was first proved in [Sz].

(b) If (HS) has only non-degenerate solutions, Morse theory in the setting of [Sz$_1$] implies the existence of 2^k geometrically distinct solutions.

(c) Theorem 7.1. generalizes also the results of [CZ], [Fr]. Let us consider some special cases :

1) (Arnold's conjecture.) If H is periodic in each variable and $\int_0^T h(t)dt = 0$ then (HS) has $2N + 1$ geometrically distinct solutions. This result was first proved in [CZ] assuming the existence of a second derivative for H.

2) (Generalized Poincaré-Birkhoff theorem.) Assume that H is periodic in $p_1, \ldots, p_N$ and that $A_\infty(t) = \text{diag } [0, B_\infty(t)]$. If $\int_0^T B_\infty(t)dt$ is invertible and if $\int_0^T h_n(t)dt = 0$, $n = 1, \ldots, N$, then (HS) has at least $N + 1$ geometrically distinct solutions. This result generalizes theorem 3 of [CZ] and contains the forced pendulum equation in Hamiltonian form.

3) (Second order systems.)   Consider the problem

$$\frac{d}{dt}(M(t)q) + \nabla V(t, q) = f(t)$$
$$q(0) - q(T) = \dot{q}(0) - \dot{q}(T) = 0 \qquad\qquad (HS')$$

where $M(t)$ is a symmetric $N \times N$ matrix depending continuously on t and such that, for some $\lambda > 0$ and all $(t, q) \in [0, T] \times \mathbb{R}^N$,

$$(M(t)q, q) \geq \lambda \mid q \mid^2 .$$

The corresponding Hamiltonian is given by

$$H(t, p, q) = \frac{1}{2}(M(t)^{-1}p, p) + V(t, q).$$

We assume that $f \in C([0, T], \mathbb{R}^N)$ and that

(V1) $\nabla V(t, q) = B_\infty(t)q + o(\mid q \mid)$

as $\mid q \mid \to \infty$, uniformly in t, where $B_\infty(t)$ is a symmetric matrix which is continuous in t.

(V2)   The space N of solutions of the periodic problem

$$\frac{d}{dt}(M(t)q) + B_\infty(t)q = 0$$

is spanned by $\{v_1, \ldots, v_k\} \subseteq \mathbb{R}^N$.

(V3) $H(t, z + v_j) = H(t, z)$, $\int_0^T (f(t), v_j)dt = 0$,

$j = 1, \ldots, k$. Under the above assumptions, problem (HS') has at least $k + 1$ geometrically distinct solutions. This result generalizes theorem 3 of [FM]. It is applicable to the Josephson multipoint system and to linearly coupled forced pendulums (see [Ma]).

8. A Perturbation Theorem

In this section we consider a connected Finsler manifold X of class C^1 together with a sequence $(X_n)_{n \geq 0}$ of closed, connected submanifolds of class C^2 of X. We assume that there exists, for every $n \geq 0$, a retraction $r_n : X \to X_n$; the limit relative category is computed with respect to (X_n).

Theorem 8.1. *Let $f \in C^1(X, \mathbb{R})$, $Y \subset Z$ closed subsets of X, $a \leq \alpha < \beta \leq b$ and $\Psi : Z \to \mathbb{R}$ be continuous. Assume that*

a) $\sup_Y f < a$, $f^a <_Y^\infty \Psi^\alpha$, $\Psi^\beta <_Y^\infty f^b$;
b) *f satisfies the $(PS)_c^*$ condition with respect to (X_n) for every $c \in [a, b]$;*
c) *There exists $\eta > 0$ such that $f^{-1}([a - \eta, b + \eta])$ is complete.*

Then $f^{-1}([a, b])$ contains at least $cat_{X,Y}^\infty(\Psi^\beta) - cat_{X,Y}^\infty(\Psi^\alpha)$ critical points of f.

Proof: Let $\dot{c}_j$ be defined as in theorem 6.1 and assume that

$$\text{cat}^\infty_{X,Y}(\Psi^\alpha) < j \le \text{cat}^\infty_{X,Y}(\Psi^\beta).$$

It follows from a) that $a \le c_j \le b$. It suffices then to apply theorem 6.1. $\square$

Theorem 8.1. is motivated by a recent result of Felmer [Fe$_1$] which generalizes theorem 6.10 in [FW$_1$]. Consider the following problem:

$$\frac{d}{dt} L_y(t, u(t), u(t)) = L_x(t, u(t), u(t))$$
$$u(0) - u(T) = u(0) - u(T) = 0 \quad (LS)$$

where $L : \mathbb{R} \times \mathbb{R}^N \times \mathbb{R}^N \to \mathbb{R}$

$$L(t, x, y) = (1/2)(M(t, x)y, y) - V(x) + (h(t), x)$$

is such that the following conditions hold:

(L1) $M(t, x)$ is a symmetric matrix of order N continuously differentiable on $[0, T] \times \mathbb{R}^N$ and such that

$$(M(t, x)y, y) \ge \lambda \mid y \mid^2$$

for some $\lambda > 0$ and every $(t, x, y) \in [0, T] \times \mathbb{R}^N \times \mathbb{R}^N$.

(L2) $V \in C^1(\mathbb{R}^N, \mathbb{R}^N)$.

(L3) $M(t, x)$ and $V(x)$ are T_n-periodic in x_n, $1 \le n \le N$.

(L4) $h \in C([0, T], \mathbb{R}^N)$ and $\int_0^T h(t)dt = 0$.

The solutions of (LS) are the critical points of the functional

$$f(u) = \int_0^T L(t, u(t), u(t))dt$$

defined on the space H^1_T. Let us recall that

$$H^1_T = \{u \in L^2(\mathbb{T}^1; \mathbb{R}^N) : u \in L^2(\mathbb{T}; \mathbb{R}^N)\}.$$

The scalar product on H^1_T is given by

$$\langle u, v \rangle = \int_0^T [(u(t), v(t)) + (u(t), v(t))]dt.$$

Since, by assumption (L3),

$$f(u + T_n e_n) = f(u), \quad 1 \le n \le N,$$

(where $e_1, \ldots, e_N$ denotes the canonical basis of $\mathbb{R}^N$) it is natural to define f on the manifold $X = E \times \mathbb{T}^N$ where E is the orthogonal space to the constant functions in H^1_T. Thus every $u \in X$ has the representation $u = \tilde{u} + \bar{u}$ where $\tilde{u} \in E$ and $\bar{u} \in \mathbb{T}^N$. By the classical Lusternik-Schnirelman theory f has at least $N + 1$ critical points (see e.g. [MW], [Ra$_1$]). We shall prove a more precise result due to Felmer [Fe$_1$]. Let us denote by ψ the restriction of f to $\mathbb{T}^N$ and by $\mathcal{V}$ the maximum of $| \nabla V(x) |$ over $\mathbb{R}^N$.

Theorem 8.2. *Under the above assumptions, let $\gamma < \alpha < \beta$ be such that*

$$m := cat_{\mathbb{T}^N, \psi^\gamma}(\psi^\beta) - cat_{\mathbb{T}^N, \psi^\gamma}(\psi^\alpha) > 0,$$
$$(T^{3/2}\mathcal{V} + T \mid h \mid_{L^2})^2 < 8\pi^2 \lambda(\alpha - \gamma).$$

Then f has at least m critical points with critical values in $[\gamma, \beta]$.

Proof: We shall apply theorem 8.1 with $X_n \equiv X, Y = \psi^\gamma$ and $Z = \mathbb{T}^N$. The space X is complete and it is proved in [MW] that f satisfies the $(PS)_c^*$ condition with respect to (X_n) for every $c \in \mathbb{R}$. (In our case, the $(PS)_c^*$ condition is nothing but the classical Palais-Smale condition at the level c.) Since E is contractible, we have

$$\text{cat}_{\mathbb{T}^N, \psi^\gamma}(\psi^\beta) = \text{cat}_{X, \psi^\gamma}(\psi^\beta), \quad \text{cat}_{\mathbb{T}^N, \psi^\gamma}(\psi^\alpha) = \text{cat}_{X, \psi^\gamma}(\psi^\alpha).$$

We choose $b = \beta$ and $a > \gamma$ such that

$$(T^{3/2}\mathcal{V} + T \mid h \mid_{L^2})^2 < 8\pi^2 \lambda(\alpha - a). \tag{1}$$

It is clear that $\psi^\beta \subseteq f^b$ and $\psi^\gamma \subseteq f^a$. It remains only to prove that $f^\alpha <_{\psi^\gamma}^\infty \psi^\alpha$. Let $u \in f^a$. Using (L1), (L4) and the Cauchy-Schwarz inequality, we obtain

$$-\int_0^T V(u)dt \leq a - \frac{\lambda}{2} \mid u \mid_{L_2}^2 + \mid h \mid_{L^2} \mid \tilde{u} \mid_{L^2}. \tag{2}$$

By the mean value theorem, the definition of $\mathcal{V}$ and the Cauchy-Schwarz inequality, we have

$$\left| \int_0^T (V(u) - V(\overline{u}))dt \right| \leq \mathcal{V} \int_0^T \mid \tilde{u} \mid dt \leq \mathcal{V}T^{1/2} \mid \tilde{u} \mid_{L^2}. \tag{3}$$

By using (2), (3) and Wirtinger inequality, we find that

$$\begin{aligned} \psi(\overline{u}) &\leq a - \frac{2\pi^2}{T^2}\lambda \mid \tilde{u} \mid_{L^2}^2 + (\mid h \mid_{L^2} + \mathcal{V}T^{1/2}) \mid \tilde{u} \mid_{L^2} \\ &\leq a + T^2(\mid h \mid_{L^2} + \mathcal{V}T^{1/2})^2/(8\pi^2\lambda). \end{aligned}$$

Inequality (1) implies that $\psi(\overline{u}) \leq \alpha$. Thus we have proved that, if $u \in f^a$, then $\overline{u} \in \psi^\alpha$. The deformation

$$[0, 1] \times f^a \to X : (t, u) \to (1 - t)\tilde{u} + \overline{u}.$$

shows that $f^a <_{\psi^\gamma}^\infty \psi^\alpha$.

References

[BB] A. Bahri, H. Berestycki, Existence of forced oscillations for some nonlinear differential equations, Comm. Pure Appl. Math., **37**(1984), 403-442.

[Be] V. Benci, A geometrical index for the group S^1 and some applications to the study of periodic solutions of ordinary differential equations, Comm. Pure Appl. Math. **34**(1981), 393-432.

[BR] V. Benci, P. Rabinowitz, Critical points for indefinite functionals, Inv. Math. **52**(1979), 241-273.

[Ch] K.C. Chang, On the periodic nonlinearity and multiplicity of solutions, Nonlinear Analysis, TMA, **13**(1989), 527-537

[CLZ] K.C. Chang, Y. Long, E. Zehnder , Forced oscillations of the triple pendulum, "Analysis, et Cetera", edited by P. Rabinowitz and E. Zehnder, Academic Press, Boston (1990), 177-208.

[CP] M. Clapp, D. Puppe, Critical point theory with symmetries, preprint, J. Reine Angew. Math. **418**(1991) 1-29.

[CZ] C. Conley, E. Zehnder, The Birkhoff-Lewis fixed point theorem and a conjecture of V. Arnold, Invent. Math. **73**(1983), 33-45.

[Do] A. Dold, "Lectures on Algebraic topology", Springer Verlag, Berlin, 1972.

[Fe$_1$] P. Felmer, Multiple solutions for Lagrangian systems in $\mathbb{T}^n$, preprint, Nonl. Anal. T.M.A. **15**(1990) 815-831.

[Fe$_2$] P. Felmer, Periodic solutions of spatially periodic hamiltonian systems, preprint, 1989.

[FM] A. Fonda, J. Mawhin, Multiple periodic solutions of conservative systems with periodic nonlinearity, "Proceedings of the International Conference on Theory and Applications of Differential Equations, vol. 1", edited by A.R. Aftabizadeh, Ohio University Press (1988), 298-304.

[FW$_1$] G. Fournier, M. Willem, Multiple solutions of the forced double pendulum equation, "Analyse Non Linéaire", Gauthier-Villars, Paris (1989), 259-281.

[FW$_2$] G. Fournier, M. Willem, Simple variational methods for unbounded potentials, in "Topological fixed point theory and applications", Springer Verlag, New York, 1989, 75-82.

[FW$_3$] G. Fournier, M. Willem, The mountain circle theorem, Lecture note in Mathematics, Springer, **1475**(1991) 147-160.

[FW$_4$] G. Fournier, M. Willem, Relative category and the calculus of variations, "Variational Methods Proceeding of a Conference, Paris, June 1988", edited by H. Berestycki, J.M. Coron, I. Ekeland, Birkh%₀user, Boston, Basel, Berlin (1990), 95-104.

[Fr] J. Franks, Generalizations of the Poincaré-Birkhoff theorem, Annals of Mathematics **128**(1988), 139-151.

[Hu] S.T. Hu, "Theory of retracts", Wayne State University Press, Detroit, 1965.

[Li] J.Q. Liu, A generalized saddle point theorem, J. Diff. Equ. **92**(1989), 372-395.

[LL] J.Q. Liu, S. Li, Some existence theorems on multiple critical points and their applications, Kexue Tongbao, Vol. 17 (1984) (in chinese).

[Ma] J. Mawhin, Forced second order conservative systems with periodic nonlinearity, "Analyse Non Linéaire", Gauthier-Villars, Paris (1989), 415-433.

[MW] J. Mawhin, M. Willem, "Critical point theory and Hamiltonian systems", Springer Verlag, New York, 1989.

[Pa] R.S. Palais, Critical point theory and the minimax principle, in "Proc. Symp. Pure Math.", Vol. 15, Amer. Math. Soc., Providence, R.I., 1970, 185-212.

[Pa$_1$] R.S. Palais, Lusternik-Schnirelman theory on Banach manifolds, Topology, 5(1966), 115-132.

[Ra] P. Rabinowitz, "Minimax methods in critical point theory with applications to differential equations", C.B.M.S. Reg. Conf. 65, Amer. Math. Soc., Providence, R.I., 1986.

[Ra$_1$] P. Rabinowitz, On a class of functionals invariant under a $\mathbb{Z}^n$ action, Trans. Amer. Math. Soc. 310(1988), 303-311.

[Re] M. Reeken, Stability of critical points under small perturbations, Part. I : Topological theory, Manuscripta Math. 7(1972), 387-411.

[Sp] E. Spanier, "Algebraic topology", McGraw Hill, New York, 1966.

[Sz] A. Szulkin, A relative category and applications to critical point theory for strongly indefinite functionals, preprint, Nonlinear Anal. T.M.A. 15(1990) 725-739.

[Sz$_1$] A. Szulkin, Cohomology and Morse theory for strongly indefinite functionals, preprint, 1990.

[Ta] G. TARANTELLO, Remarks on forced equations of the double pendulum type, Trans Amer. Math. Soc. 326(1991) 441-452.

[Ta$_1$] G. TARANTELLO, Multiple forced oscillations for the N-pendulum equation, Comm. Math. Phys. (1990) 499-517.

[Wi] M. WILLEM, "Lectures on critical point theory", Trabalho de Mat., 199,Fundação Univ. Brasilia, Brasilia, 1983.

Coexistence of Infinitely Many Stable Solutions to Reaction Diffusion Systems in the Singular Limit

Yasumasa Nishiura

Division of Mathematics and Informatics, Faculty of Integrated Arts and Sciences, Hiroshima University, Higashi-Hiroshima 724, Japan

1. Introduction and singular limit slow dynamics

Recognition stems from realization of the separation boundary between two different physical or chemical *states*. In other words we can observe natural phenomena through the emergence and evolution of the *interface* between these states as in solidification, combustion, chemical reaction, and biological patterns. The interface studied here results from the balance between two opposing tendencies: a *diffusive* effect and a (physical or chemical) *separation kinetics* built in the system. The former attempts to smooth out the inhomogeneity as in the heat equation, and the latter drives the system to one or the other pure state such as solid or liquid (see, for instance, Fife [19] for details). Turing's contribution [58] is one of the pioneering works related to the onset of spatial patterns through a cooperative work of diffusion and separation kinetics. Besides the existence of these two tendencies, another key ingredient to produce interesting interfacial patterns is the differences of the strength of the above mixing and unmixing effects among species involved in the system. In fact, reaction diffusion systems for two components u and v, which are the main concern in this paper, can be classified formally as

- (i) There is a difference in the diffusion rates of u and v;
- (ii) There is a difference in the reaction rates of u and v;
- (ii) There are differences in the diffusion and reaction rates of u and v,
  i.e., a combination of (i) and (ii).

Steady interfacial patterns, which usually originate in the onset of symmetry breaking patterns through Turing's diffusion driven instability, are commonly observed in the first category. A typical example is an activator-inhibitor system describing morphogenetic patterns (see Meinhardt[38] and Murray [40; chapters 14 and 15]). Propagator-controller systems, including a simple skeleton model for the Belousov-Zhabotinsky reaction, lie in the second category (see, for instance, Fife [18], and Keener and Tyson [33]). Layer oscillation ("breather") is one of the characteristic phenomena in the third category (see Nishiura and Mimura [46]).

In this paper we focus on the first category and consider the following system in one-dimensional space:

$$\delta u_t = \varepsilon^2 u_{xx} + f(u, v) \tag{1.1a}$$

$$\text{in } I$$

$$v_t = D v_{xx} + g(u, v) \tag{1.1b}$$

$$u_x = 0 = v_x \qquad \text{on } \partial I, \tag{1.1c}$$

where I is the unit interval $(0, 1)$, δ denotes the ratio of reaction rates of u and v, and ε^2 and D are the diffusion coefficients of u and v, respectively. We assume that $0 < \varepsilon << 1$ but $D = O(1)$, i.e., (1.1) belongs to the first category.

Although $\delta = O(1)$ is a typical situation in this category, the main results hold for the wider regime where $\varepsilon/\delta = o(1)$ as $\varepsilon \downarrow 0$, in particular, one can take $\delta = \varepsilon^\alpha$ with $0 \leq \alpha < 1$ (see Remark 2.8). The nullclines for typical f and g are drawn in Figure 1.1; $f = 0$ is of sigmoidal shape; $g = 0$ intersects with $f = 0$ transversally, and $f > 0$, $g > 0$ in lower regions of those nullclines. More precise assumptions will be stated at the end of this section.

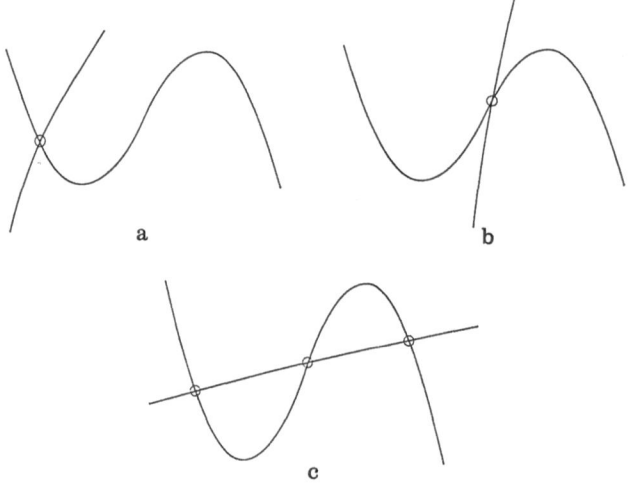

Figure 1.1.

One of the most important features of the nonlinearity is the bistable nature of f. Namely, for a fixed v, the kinetics $u_t = f(u, v)$ drives u toward either the left-end zero or the right-end zero of $f = 0$ depending on the initial data. This separating force causes the emergence of transition layers when the initial data is distributed in spatial direction (see Figure 1.2(a)). It turns out that the width of the resulting layers becomes $O(\varepsilon)$. Then these layers start to propagate slowly ($O(\varepsilon/\delta)$-speed) in some direction, but this wave can be blocked at some stage by the inhibitor v, and settles down to a steady state (see Figure 1.2(b)), since v can diffuse much faster than u and make an environment that controls the motion of u. Note that the ratio of unit time scales of Figures (a) and (b) is $3 : 40$, i.e., layers propagate very slowly. Our main concern is the stability of such a layered solution obtained as the final pattern in Figure 1.2(b).

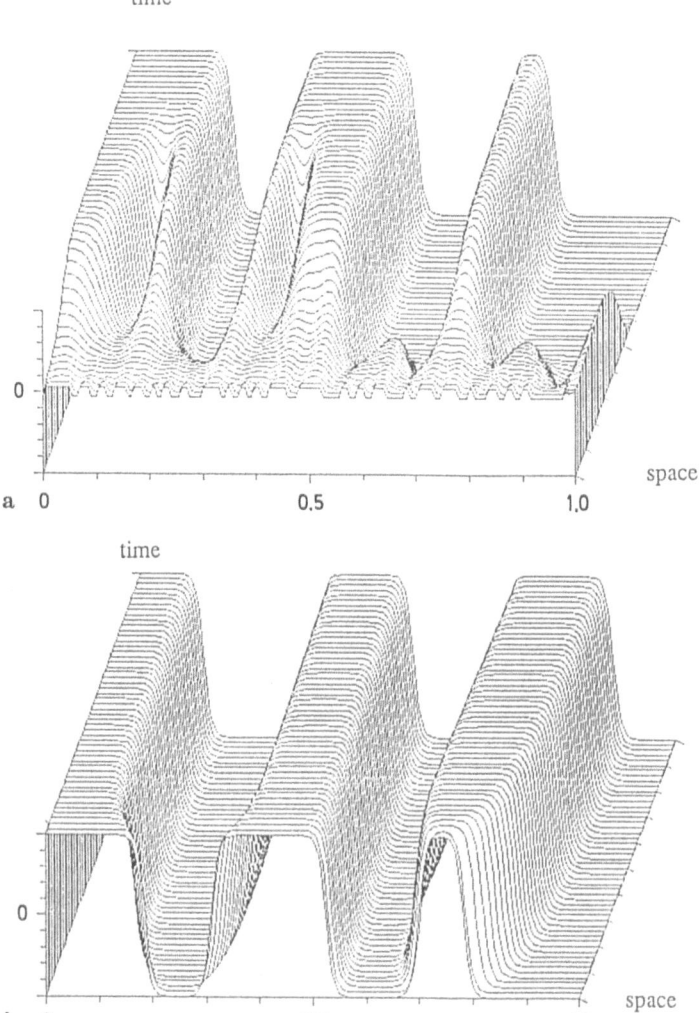

Figure 1.2.

The basic question is " How many such stable layered solutions and what do they look like? " The answer is, which is our final goal, is really remarkable, namely,

Main Theorem. *The number of non-constant stable steady states of* (1.1) *becomes "infinite" in the singular limit* $\varepsilon \downarrow 0$. *In other words, for any large number N, one can find an* ε_0 *such that* (1.1) *has at least N asymptotically stable steady states for* $0 < \varepsilon \leq \varepsilon_0$.

In fact we will prove in Section 3 that for an arbitrary number n, the *normal n-layered solution* (see Corollary 3.9) as in Figure 1.3 becomes stable for small ε. Throughout the paper we use the word "stable" in the sense of "asymptotically stable" in an appropriate norm.

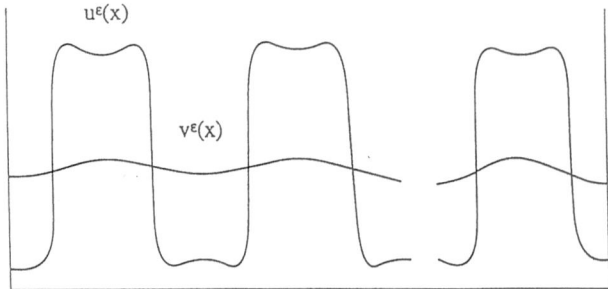

Figure 1.3.

Apparently this makes a sharp contrast with a scalar reaction diffusion equation.

$$\delta u_t = \varepsilon^2 u_{xx} + f(u), \qquad u_x = 0 \text{ on } \partial I, \qquad (1.2)$$

where f is typically a cubic-like function of u. It is known (Casten and Holland [10], and Matano [37]) that any non-constant solution of (1.2) is, if it exists, *unstable*. If some constraint (for instance, mass conservation) is imposed on (1.2), then (1.2) may have a unique (up to reflection) non-constant stable solution with one internal transition layer (see Carr, Gurtin, and Slemrod [8]). However, all the remaining multi-layered solutions are unstable. A natural question is "What is the mechanism that causes the above difference between scalar and system?" It is obvious that the second component v somehow controls the behavior of u, but in what manner ?

In order to see the role of the controller v more clearly, we shall derive a *singular limit slow dynamics* from (1.1) in a heuristic way. Hereafter we assume for definitness that (f, g) is of type (b) in Figure 1.1. The dynamics of (1.1) consists of two stages with different time scales; outer dynamics (phase separation process) and then followed by the slow layer dynamics (propagation process). Let $(u_0(x), v_0(x))$ be a smooth and moderate initial data, then the diffusion term $\varepsilon^2 u_{xx}$ could be neglected for a while until u_{xx} becomes sufficiently large. The resulting *outer dynamics* is

$$\begin{aligned} \delta U_t &= f(U, V) \\ V_t &= DV_{xx} + g(U, V) \end{aligned} \qquad (1.3a)$$

with

$$V_x = 0 \qquad \text{on } \partial I. \qquad (1.3b)$$

If layers move slowly compared with the relaxation time of (1.3) (see (1.11)), we can expect that the solution of (1.3) approaches a steady state in regions away from layers:

$$\begin{aligned} 0 &= f(U, V) \\ 0 &= DV_{xx} + g(U, V) \end{aligned} \qquad (1.4)$$

Since f is of bistable type for a fixed V as in Figure 1.1, U is attracted to either $u = h_-(v)$ or $u = h_+(v)$ branch everywhere except in neighbourhoods of finitely many points $\{\varphi_i\}_{i=1}^n$ for generic data. Hence, to the lowest degree of approximation, (1.4) can be rewritten as

$$0 = DV_{xx} + G_\Phi(V) \tag{1.5}$$

with $(1.3)_b$, where Φ denotes a collection of layer positions $\{\varphi_i\}_{i=1}^n$ ($0 < \varphi_1 < \varphi_2 < \cdots < \varphi_n < 1$), and $G_\Phi(V)$ is equal to either $G_-(V) \equiv g(h_-(V), V)$ or $G_+(V) \equiv g(h_+(V), V)$, according to the chosen branch, on each subinterval partitioned by $\{\varphi_i\}_{i=1}^n$. $G_+(V)$ (resp. $G_-(V)$) is defined for $v < \bar{v}$ (resp. $v > \underline{v}$), positive (resp. negative), and strictly decreasing (see Figure 1.4 and Remark 1.1).

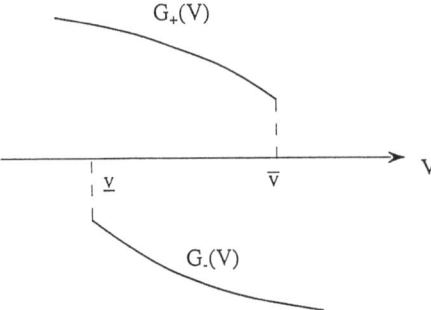

Figure 1.4.

Since G_Φ has discontinuities at layer positions, we say that V is a solution of (1.5) if and only if it satisfies (1.5) in a classical sense in each subinterval as well as (1.3b), and it is matched in C^1-sense at each layer position, i.e., the right and left limits coincide each other up to first derivatives. The associated U-component has jump discontinuities at φ_i's where two stable branches are switched. Note that (1.5), in general, does *not* have a solution for arbitrary partition Φ. In fact, even for mono-layer case ($n = 1$), φ_1 cannot be arbitrary close to boundary points 0 and 1, because $|G_\pm(V)|$ are bounded away from zero (see Theorem 2.1). On the other hand, if (f, g) is of bistable type (Figure 1.1 (c)), $G_\pm$ has a unique zero respectively, and hence φ_1 can be taken arbitrarily.

After the completion of outer dynamics, we move into the next stage: the process of layer propagation. The diffusion term $\varepsilon^2 u_{xx}$ plays an important role. Once (u, v) approaches a solution of (1.4) in outer region, it is supposed to be held rigidly there, since we assume that layers propagate slowly. Therefore we can localize our analysis in the neighbourhood of each layer position to derive the propagation dynamics. We shall introduce the following stretched coordinate and slow time scale to study the dynamics of u inside of thin layers.

Let φ be an arbitrary layer position and define a *stretched* coordinate y and a *slow time* s:

$$y \equiv \frac{x - \varphi}{\varepsilon}, \qquad s \equiv \frac{\varepsilon}{\delta} t. \tag{1.6}$$

The κ-neighbourhood $I_\kappa \equiv (\varphi - \kappa, \varphi + \kappa)$ of $x = \varphi$ is stretched to $\tilde{I}_\kappa \equiv (-\kappa/\varepsilon, \kappa/\varepsilon)$. What we want to know is the time dependence of φ, and it turns out that φ becomes a function of slow time s. In fact, noting the relation,

$$\frac{\partial}{\partial t} = \frac{\varepsilon}{\delta}\frac{\partial}{\partial s} - \frac{1}{\varepsilon}\frac{d\varphi}{dt}\frac{\partial}{\partial y} \qquad \text{and} \qquad \frac{\partial}{\partial x} = \frac{1}{\varepsilon}\frac{\partial}{\partial y},$$

the first equation of (1.1) becomes

$$\varepsilon u_s - \frac{\delta}{\varepsilon}\varphi_t u_y = u_{yy} + f(u, V(\varphi + \varepsilon y)). \tag{1.7a}$$

Here v is replaced by the solution V of (1.5). In view of the second term of the left-hand side of (1.7a), we see that it is natural to regard φ to be a function of s instead of t. Then (1.7b) becomes

$$\varepsilon u_s - \varphi_s u_y = u_{yy} + f(u, V(\varphi + \varepsilon y)). \tag{1.7b}$$

Taking a formal limit of (1.7b) as $\varepsilon \downarrow 0$, we have

$$u_{yy} + \varphi_s u_y + f(u, V(\varphi)) = 0 \qquad \text{on } \mathbf{R}. \tag{1.8a}$$

The stretched interval $\tilde{I}_\kappa$ becomes a whole line $\mathbf{R}$ in this limit and the boundary conditions become

$$u(\pm\infty) = h_\pm(V(\varphi)) \qquad (\text{ resp. } h_\mp(V(\varphi)) \tag{1.8b}$$

if the outer solution to the right of $x = \varphi$ is attracted to the branch $u = h_+(V)$ (resp. $h_-(V)$). Hereafter we focus on the former case (h_+-case). It is well-known (see, for instance, Fife and McLeod [22] and references therein) that (1.8) has a unique solution $u = u(y; V(\varphi))$ provided that

$$\varphi_s = c(V(\varphi)) \tag{1.9}$$

holds, where $c(\cdot)$ is, what is called, the *velocity function* of traveling waves for the bistable nonlinearity f. Typically c is a strictly monotone increasing function of V and has a unique zero at $v = v^*$ (see (A.2)). In fact, when f is given by

$$f(u, V) = u(1 - u)(u - V), \tag{1.10a}$$

the solution and its velocity function with $u(-\infty) = 0$ and $u(\infty) = 1$ are uniquely determined as

$$u(y) = \frac{1}{2} + \frac{1}{2}\tanh\left(\frac{y}{2\sqrt{2}}\right) \tag{1.10b}$$

$$c(V) = \sqrt{2}(V - \frac{1}{2}). \tag{1.10c}$$

In terms of original time scale t, (1.9) becomes

$$\varphi_t = \frac{\varepsilon}{\delta}c(V(\varphi)), \tag{1.11}$$

which shows that each internal layer moves slowly compared with the relaxation time of the outer dynamics (1.3) so long as $\varepsilon/\delta = o(1)$ as $\varepsilon \downarrow 0$. Note that, when $\delta = O(\varepsilon)$, the motion of φ is not slow for small ε, hence the associated dynamics becomes different from (1.12) below. See the discussion in Section 5.

Summarizing the above discussion, the slow dynamics for n layers, to the lowest degree of approximation, is given by

$$(\varphi_i)_s = (-1)^{i-1} c(V(\varphi_i(s))) \tag{1.12a}$$

$$DV_{xx} + G_{\Phi_n(s)}(V) = 0 \tag{1.12b}$$

with the ordering property

$$0 < \varphi_1(s) < \varphi_2(s) < \cdots < \varphi_n(s) < 1, \tag{1.12c}$$

where $\Phi_n(s) = \{\varphi_1(s), \varphi_2(s), \cdots, \varphi_n(s)\}$ denotes the locations of n layers. Here the coefficient $(-1)^{i-1}$ of (1.12a) comes from the assumption that $G_{\Phi_n(s)}(V)$ is equal to $G_-(V)$ on the first subinterval $(0, \varphi_1(s))$. Note that (1.12b) has a *unique* solution for a given $\Phi_n(s)$ because of (1.22).

As far as the number of layers remains unchanged, (1.12) can be regarded as a system of nonlinear ODEs with respect to $(\varphi_1, \varphi_2, \cdots, \varphi_n)$. However (1.12) has several quite different features from usual ODE systems. Firstly the definition domain for $\Phi_n = (\varphi_1, \cdots, \varphi_n)$ is not a priori clear since the vector field $(1.12)_a$ is defined only on Φ_n where (1.12b) is satisfied. In fact, as was remarked earlier, the solution of (1.12) does not always exist for an arbitrary Φ_n with (1.12c). However it turns out in Section 2 that (1.12) is well-defined for any type of nonlinearity in Figure 1.1 at least in a neighbourhood of a critical point Φ_n^* defined later on, which is sufficient for the study of local stability of it. Secondly the number of unknowns may decrease when time evolves. Namely, two layers may collide with each other and *disappear* after that. This reminds us what is called the *coarsening* process in solidification theory. In Figure 1.5 there are four layers initially, however, after a finite time, two layers collide and disappear, then approach two-layered solution. Figure 1.5(a) shows the orbits of (1.12) with $n = 4$, and Figure 1.5(b) shows the u-profile of the solution to the original system (1.1) for the corresponding initial data.

Despite this singularity, the solution is well-defined even at collision points, and can be continuated after that, since the C^1-matching conditions do not break down at the hitting time. Taking into account this reduction of layer number, we see that, when we start with an *N-layered* solution, more natural definition domain for Φ_n is given by

$$\hat{\mathcal{M}}_N = \bigcup_{n=1}^{N} \mathcal{M}_n, \tag{1.13}$$

which is finite dimensional, where $\mathcal{M}_n$ is defined by

$$\mathcal{M}_n = \{\Phi_n| \text{ there exists a solution of (1.12b) for } \Phi_n \}. \tag{1.14}$$

The whole space for (1.12) is apparently given by

$$\hat{\mathcal{M}}_\infty = \bigcup_{n=1}^{\infty} \mathcal{M}_n \tag{1.15}$$

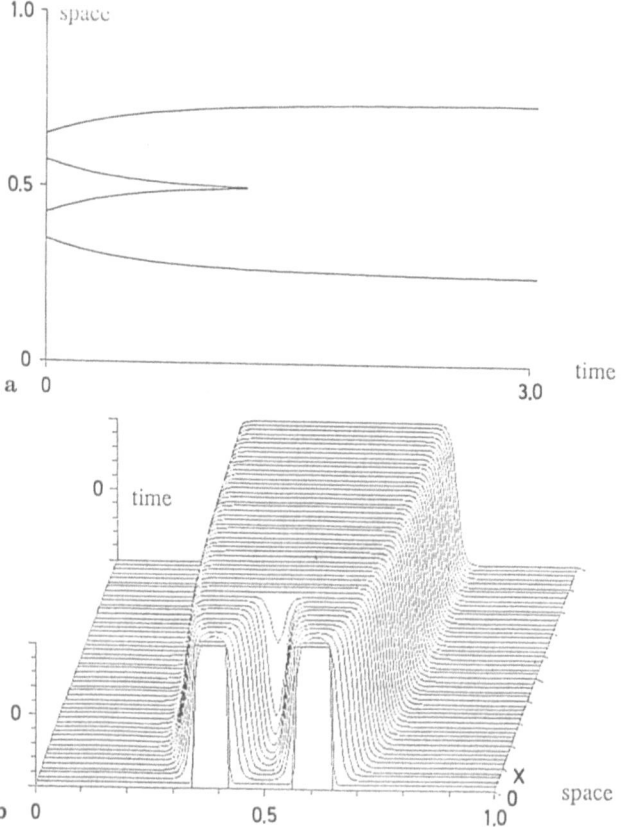

Figure 1.5.

which is an *infinite* dimensional space. Sometimes it is more convenient to consider the pair $(\Phi_n, v(\Phi))$ instead of only Φ_n where $v(\Phi)$ denotes the n-dimensional vector $(V(\varphi_1), V(\varphi_2), \cdots, V(\varphi_n))$, since the motion of each layer is determined by the value of V there. We use the same notation as before for this new definition such as

$$\mathcal{M}_n = \{(\Phi_n, v(\Phi_n)) | \text{ there exists a solution of } (1.12)_b \text{ for } \Phi_n \}. \tag{1.16}$$

We call $\mathcal{M}_n$ the *slow manifold for n-layered solutions* and $\hat{\mathcal{M}}_\infty$ the slow manifold for (1.1). It may not be appropriate to call $\hat{\mathcal{M}}_N$, $\hat{\mathcal{M}}_\infty$ "manifold", since they are the union of manifolds of different dimension. However we abuse this terminology to call these objects. Note that the velocity of each layer is determined locally by $(1.12)_a$, however $\varphi_i(s)$ have strong linkage with each other through the relation $(1.12)_b$ for the controller V, which is a nonlocal relation of them.

It should be noted that, for a given number of layers, there is a *unique* critical point for (1.12). In fact suppose that $\Phi_n^* = (\varphi_1^*, \cdots, \varphi_n^*)$ is a critical point, then $V(\varphi_i^*) = v^*$ because of $c(v^*) = 0$. By using a phase plane analysis for $(1.12)_b$, which is glued at $V = v^*$, and the assumption $\left(\dfrac{dG_\pm}{dV} < 0 \right)$, we can prove without difficulty that there is a unique orbit that rotates around $(v^*, 0)$ in the phase plane $[\frac{n}{2}]$ (resp. $[\frac{n}{2}]$ and half) times depending on n being even (resp. odd). We call this unique critical point Φ_n^* the *normal n-layered solution for the singular limit slow dynamics* (1.12). The normal layered solutions are important, since they seem to form the attractor of (1.12).

Conjecture. *After the coarsening process, any solution of (1.12) approaches one of the normal layered solutions $\{\Phi_n^*\}_{n=1}^\infty$.*

When the number of initial layers is less than or equal to 2, this is true (see Nishiura and Suzuki [48]).

Now we are ready to answer the question concerning the role of controller v. First we consider the scalar equation (1.2). Since there are no controllers in this case, the associated slow dynamics consists of *only* (1.12a) with $V(\varphi_i(s))$ being equal to some fixed value ξ. Recalling (1.10b), we easily see that internal layers cannot persist as a steady state, since they steadily move with constant velocity unless $\xi = 0$. Even for $\xi = 0$, they are *neutrally stable*, because arbitrary positions are equilibrium points of (1.12a). Recent progress on slow motions or metastable patterns (see Carr and Pego [9], Fusco and Hale [26], and Alikakos, Bates, and Fusco [2]) which gives us a more accurate approximation than (1.12a), shows that, when $\xi = 0$, layers are not neutrally stable but move with transcendentally small velocities. In any case layer structure cannot persist without a controller V. Then the question is how the controller V stabilizes the layer structure. The key for this lies in the slow manifold $\mathcal{M}_n$ on which interfaces move around. In order to see the role of the controller and slow manifold more clearly, we digress a little bit from (1.12) and consider intuitively how we can stabilize a mono-layered solution (Figure 1.6 (a)) of the scalar equation

$$\delta u_t = \varepsilon^2 u_{xx} + f(u, V) \tag{1.17a}$$

with f being given by (1.10a). Note that $v^* = 1/2$ in this case.

A naive way to stabilize this layer is to control the area $A = \int_I u \, dx$ so that it approaches neither 0 and 1. To do this, we add the following auxiliary equation to the scalar equation (1.17a) so that the *scalar* controller $V \equiv \xi$ steers u toward an internal layer solution with the assigned area A with $0 < A < 1$.

$$\frac{d\xi}{dt} = \left(\int_I u \, dx - A \right) - (\xi - \frac{1}{2}), \tag{1.17b}$$

Similar arguments to derive the slow dynamics (1.12) also work for this system (1.17), and the resulting one is given by

$$\varphi_s = c(\xi) \tag{1.18a}$$

$$\frac{3}{2} - \varphi - A - \xi = 0. \tag{1.18b}$$

Since the area of the corresponding u-profile is equal to $1 - \varphi$, (1.18b) is easily obtained by computing the right-hand side of (1.17b). One can regard the scalar relation (1.18b) to be a slow manifold $\mathcal{M}$ in (φ, ξ)-space. Apparently (1.18) is equivalent to the scalar

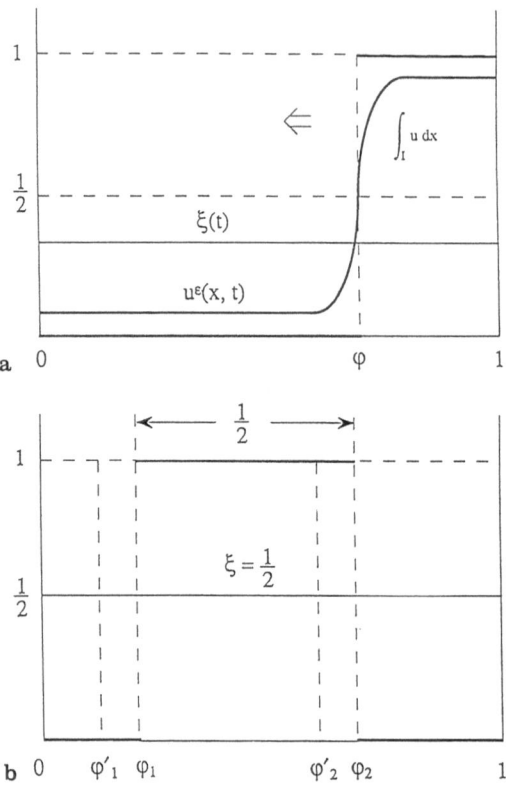

Figure 1.6.

ODE:

$$\varphi_s = c\left(\frac{3}{2} - \varphi - A\right).$$

(1.19)

Since the velocity function c is strictly monotone increasing, we see from (1.19) and $v^* = 1/2$ that $\varphi^* \equiv 1 - A$ is a unique critical point which is *globally asymptotically stable on* $\mathcal{M}$. Note that the area of the corresponding u-profile is equal to A as we expected. Although the above introduction of (1.17b) looks very artificial, it is possible to derive it in a natural way from (1,12). In fact, when D becomes sufficiently large, we can show that V tends to be flat (see [41] and [30]), i.e., a constant function in spacial direction, which is denoted by $V = \xi(s)$. On the other hand $\int_I G_{\Phi(s)}(V)dx = 0$ always holds because of Neumann boundary conditions. We, therefore, have the following relation

$$\int_I G_{\Phi(s)}(\xi)dx = 0$$

(1.20)

in the limit of $D \uparrow \infty$ instead of (1.12b). It is clear that, when $g(u, v) = u - (v - 1/2) - A$, which is one of the typical cases, (1.20) coincides with (1.18b).

The above discussion is easily generalized to multi-layered solution; for instance, if there are two layeres as in Figure 1.6 (b), then the associated slow dynamics is given by

$$(\varphi_1)_s = c(\xi)$$
$$(\varphi_2)_s = -c(\xi)$$
$$0 = (\varphi_2 - \varphi_1) - \left(\xi - \frac{1}{2}\right) - A.$$
(1.21)

Apparently $(\varphi_1, \varphi_2, \xi) = \left(1 - \frac{A}{2}, 1 + \frac{A}{2}, \frac{1}{2}\right)$ is a unique equilibrium point of (1.21), which corresponds to double-layered solution. However, this is *not* stable, since any translate of it with keeping $\varphi_2 - \varphi_1 = A$ and $\xi = 1/2$ is again an equilibrium point. Namely there exists a continuum of steady states for (1.21). This suggests that *scalar* controller ξ is *not* sufficient to stabilize *two* layers simultaneously. One may guess that the controller should have n degrees of freedom in order to control n layers. This is true, in fact, if we add another appropriate scalar variable η to (1.21) which controlls the sum of φ_1 and φ_2 (the difference is already controlled by ξ), then the resulting system has a unique asymptotically stable double-layered solution. In view of the slow dynamics (1.12), the controller V is a function of x (i.e., it has *infinite* degrees of freedom), hence V has a potentiality to control arbitrary many layers. In fact, we will see in Sections 2 that all critical points $\{\Phi_n^*\}$ of (1.12) are stable at least locally (although the discussions in Section 2 is restricted to the double-layer case, the generalization to n-layer case is straightforward by using the results in Section 3.).

Summarizing the above discussions, we can say that the slow manifold $\mathcal{M}_n$ forms a field for Φ where the unique equilibrium Φ_n^* sits at the bottom of the local basin.

Thus the singular limit system (1.12) admits the *coexistence of infinitely many stable solutions simultaneously*. However, when ε becomes positive, this is no longer true, in fact, the number of the steady states of (1.1) becomes finite for a fixed $\varepsilon > 0$, and so is the number of stable ones, although it goes to infinity as $\varepsilon \downarrow 0$. This reduction of number is caused by the diffusion effect $\varepsilon^2 u_{xx}$ which reduces the precision of discrimination of different layers. To see more clearly, we consider the scalar reaction diffusion equation of bistable type on **R**

$$\delta u_t = \varepsilon^2 u_{xx} + f(u),$$

where f is a cubic-like function such as (1.10a) with V being fixed to be a constant. Note that this simplification is plausible in the following discussion, since the controller v is close to a constant in a small neighborhood of layer position. Set an initial data $u_0(x)$ which has several layers within $O(\varepsilon)$-distance like Figure 1.7.

As time proceeds, these layers merge into a *mono*-layered travelling front (see, for example, Fife and McLeod [22]). On the other hand, if the mutual distance of layers are larger than ε, say $O(\varepsilon|\log \varepsilon|)$, these, in general, stay apart. This suggests that the resolution of layers for the original system (1.1) is proportional to the size of ε.

The arguments so far have been mainly focused on the singular limit system (1.12) and the dynamics on its slow manifold.

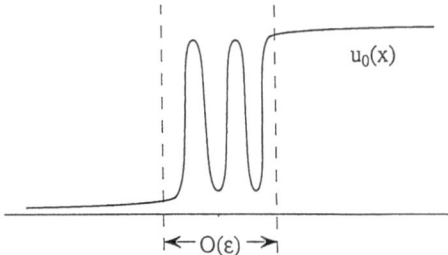

Figure 1.7.

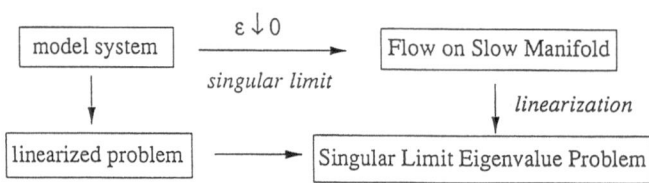

Figure 1.8.

However, it is, of course, not obvious that how the dynamics of (1.12) is related to that of the original system (1.1) for small ε. As far as stability is concerned, the Singular Limit Eigenvalue Problem (*SLEP*) method originated in Nishiura and Fujii [44] gives us a satisfactory answer, in fact, we have a commutative diagram as in Figure 1.8. Namely, a formal linearization of the singular limit slow dynamics at Φ_n^* gives us a rigorous result about the stability of normal n-layered solution to (1.1) for small ε. This not only supports the validity of the limiting slow dynamics concerning local flows near equilibrium solutions, but also is quite useful in practical sense, i.e., *formal* stability analysis of the limiting dynamics gives you a *correct* answer for the original system (1.1). We shall illustrate this more precisely in Section 2. It turns out that perturbation in outer region decays quickly, however perturbation at layer positions, essentially related to *shifting* the location of layers, behaves more slowly and delicately, and hence needs more precise analysis. Loosely speaking, the singular limit slow dynamics essentially describe the locus of each fully developped layers after outer part settles down. Accordingly, the linearized spectra at a layered solution to (1.1) is divided into two parts: *noncritical* and *critical* eigenvalues, where critical ones tend to zero as $\varepsilon \downarrow 0$ and the stability is determined by the behavior of finitely many critical eigenvalues, the number of which is proportional to that of layers. The SLEP system corresponds to the limiting eigenvalue problem for those dangerous perturbations concentrated at layer positions. Moreover we will show in Section 2 that the *formal* linearization of the limiting slow dynamics coincides exactly with the SLEP system.

The SLEP method is very close, in sprit, to the Lyapunov-Schmidt method in bifurcation theory in the sense that it enables us to reduce the linearized problem of (1.1), which is infinite dimensional, to the finite dimensional problem the size of which is proportional to the number of layers. In other words, the entire problem is contracted to the one on layers. In higher space dimension $\mathbb{R}^n$, the problem can be similarly reduced to the one on the interface, which is a $(n-1)$-dimensional hypersurface in $\mathbb{R}^n$, however it is no more a finite dimensional problem. In fact it becomes a PDE problem associated with infinitely many number of critical eigenfunctions. We shall discuss more about this in Section 5.

Apparently the linearized eigenvalue problem of (1.1) at a layered solution (see (3.12)) degenerates in a singular way and its coefficients have discontinuities at layer positions as $\varepsilon \downarrow 0$. Moreover the associated eigenfunctions with critical eigenvalues, which control the stability properties and bifurcation, do *not* remain in a usual function space, say L^2, when $\varepsilon \downarrow 0$. Technically this is the main obstacle to overcome. The basic idea of the SLEP method lies in that those dangerous critical eigenfunctions can be characterized as *distributions* by means of appropriate ε-scaling in the limit of $\varepsilon \downarrow 0$. Especially, in one-dimensional case, they become a combination of Dirac's point mass distribution on layer positions. The linearized eigenvalue problem is well-defined up to $\varepsilon = 0$ by this characterization. Also the idea of the SLEP method is free from the forms of nonlinearities, boundary conditions, and the space dimension (see Section 5).

There are several different approaches for the stability problems of large amplitude, especially, singularly perturbed solutions: One is the stability index developed by Alexander, Gardner, and Jones (see [1] and [27]). Making use of a topological approach, they presented a beautiful framework of counting the number of critical eigenvalues for a general class of systems in one-dimensional space, however their method seems inadequate to keep track of the asymptotic behavior of the critical eigenvalues. It should be noted that their approach is closely related to ours when the parameters of the system belong to the regime of singular perturbation setting (see Suzuki, Nishiura, and Ikeda [56]). Another nice work was done by Hale and Sakamoto [28] who showed existence and stability simultaneously for the inhomogeneous scalar equation, i.e., $f = f(u, x)$ in (1.2). Then, Sakamoto [55] extended this to the system case based on the results of Nishiura and Fujii [45].

Now we state the assumptions for f and g (Figure 1.9).

(A.0) f and g are smooth functions of u and v defined on some open set $\mathcal{O}$ in $\mathbf{R}^2$.

(A.1a) The nullcline of f is sigmoidal and consists of three smooth curves $u = h_-(v)$, $h_0(v)$ and $h_+(v)$ defined on the intervals I_-, I_0, and I_+, respectively. Let $\min I_- = \underline{v}$ and $\max I_+ = \bar{v}$, then the inequality $h_-(v) < h_0(v) < h_+(v)$ holds for $v \in I^* \equiv (\underline{v}, \bar{v})$ and $h_+(v)$ (resp. $h_-(v)$) coincides with $h_0(v)$ at only one point $v = \bar{v}$ (resp. $\underline{v}$) respectively.

(A.1b) The nullcline of g intersects with that of f at one or three points transversally as in Fig.1.9. The critical point on $u = h_-(v)$ (resp. $h_+(v)$ or $h_0(v)$), if exists, is denoted by $P = (u_-, v_-) = (h_-(v_-), v_-)$ (resp. $Q = (u_+, v_+) = (h_+(v_+), v_+)$ or $R = (u_0, v_0) = (h_0(v_0), v_0)$).

(A.2) $J(v)$ has an isolated zero at $v = v^* \in I^*$ such that $dJ/dv < 0$ at $v = v^*$, where
$$J(v) = \int_{h_-(v)}^{h_+(v)} f(s, v)ds.$$ Moreover we assume that $v_- < v^* < v_+$.

(A.3) $f_u < 0$ on $\mathcal{H}_+ \cup \mathcal{H}_-$, where $\mathcal{H}_-$ (resp. $\mathcal{H}_+$) denotes the part of the curve $u = h_-(v)$ (resp. $h_+(v)$) defined by $\mathcal{H}_-$ (resp. $\mathcal{H}_+$) $= \{(u, v)|u = h_-(v)$ (resp. $h_+(v))$ for $v_- \leq v < v^*(v^* < v \leq v_+)\}$, respectively. Note that v_- (resp. v_+) is replaced by $\underline{v}$ (resp. $\bar{v}$) when there are no critical points on the branch $u = h_-(v)$ (resp. $h_+(v)$). See thick solid part of $f = 0$ in Figure 1.9.

(A.4a)
$$g|_{\mathcal{H}_-} < 0 < g|_{\mathcal{H}_+}$$

(A.4b)
$$\det \left(\frac{\partial(f, g)}{\partial(u, v)} \right) \Big|_{\mathcal{H}_+ \cup \mathcal{H}_-} > 0.$$

(A.5)
$$g_v|_{\mathcal{H}_+ \cup \mathcal{H}_-} \leq 0.$$

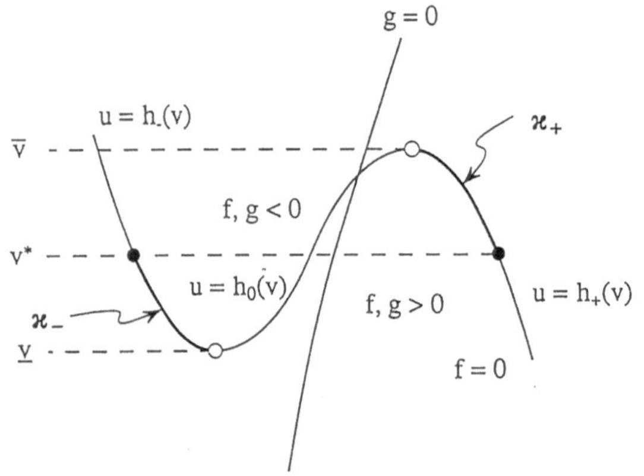

Figure 1.9.

Remark 1.1. *Let $G_\pm(v) \equiv g(h_\pm(v), v)$ for $v \in I_\pm$. Then, the assumption (A.4)(b) is equivalent to*

$$\frac{d}{dv} G_\pm(v) \Big|_{\mathcal{H}_\pm} < 0, \ respectively, \tag{1.22}$$

since it follows from $f(h_\pm(v), v) = 0$ and (A.3) that

$$\frac{d}{dv} G_\pm(v) \Big|_{\mathcal{H}_\pm} = \frac{f_u g_v - f_v g_u}{f_u} \Big|_{\mathcal{H}_\pm}.$$

Remark 1.2. *It holds that $f_u = 0$ at $(h_+(\bar{v}), \bar{v})$ and $(h_-(\underline{v}), \underline{v})$.*

The outline of this paper is as follows. In Section 2, we study the slow dynamics from a geometrical point of view and show that formal linearized eigenvalue problem of (1.12) at an equilibrium point is exactly the same as the SLEP system in Section 3 derived from the original linearized problem for (1.1). In Section 3, we prove the stability of multi-layered solutions for the original model system (1.1), and show that how the SLEP method is used to reduce the whole problem to a finite dimensional one. It turns out that the resulting SLEP system is equivalent to solving the eigenvalue problem of a tri-diagonal symmetric matrix. The signs of eigenvalues of this matrix determine the stability, and lead us to the Main Theorem. In Section 4, we explain through the intermediary of the shadow system (4.1) why the stability of layered solutions makes a sharp contrast between the system (1.1) and the single equation (1.2). In Section 5, we briefly discuss about the case of different scaling for reaction and diffusion rates (the third category mentioned at the beginning of this section), the higher dimensional problems, and several related topics for which the SLEP method is useful.

We use the following notation throughout the paper:

$C^p(\bar{I})$ = the space of p-times continuous differentiable functions on $\bar{I}$ with usual supremum norm.

$C_\varepsilon^p(\bar{I})$ = the space of p-times continuous differentiable functions on $\bar{I}$ with the norm

$$\| u \|_{C_\varepsilon^p} = \sum_{k=0}^{p} \max \left| \left(\varepsilon \frac{d}{dx} \right)^k u(x) \right|,$$

$H^p(I)$ = the usual Sobolev space of order $p(\geq 0)$ in $L^2(I)$-framework,

$H_N^p(I)$ = the space of closure of $\{\cos(n\pi x/|I|)\}_{n=0}^{\infty}$ in $H^p(I)$,

$H_0^p(I)$ = the space of closure of $\{\sin(n\pi x/|I|)\}_{n=1}^{\infty}$ in $H^p(I)$,

$H^{-1}(I)$ = the dual space of $H_N^1(I)$,

$< \cdot, \cdot >$ = the inner product in $L^2(I)$-space,

$C_{c.u.}^k(I)$-topology = the compact uniform convergence in C^k-sense in I, namely, the uniform convergence on any compact subset of I in C^k-sense.

Acknowledgments I would like to express my gratitude to Hiroshi Fujii and Masayasu Mimura for the pleasant collaboration and useful comments. Special thanks go to Hiromasa Suzuki for making numerical simulations and reading the manuscript carefully.

2. Intuitive Approach to the Stability of Multi-layered Solutions

— Slow Manifold and Formal Linearization —

2.1. Slow Manifold for Mono-layered Solution

As we observed in Section 1, the singular limit slow dynamics is finite-dimensional when we *fix* a number of layers, and the dimension of it is exactly equal to the number of layers. In this section we shall construct the slow manifolds for single and double layered solutions, respectively, and study the flow on each manifold. Let us begin with the mono-layer case (see (1.12)):

$$(\varphi)_s = c(V(\varphi(s))) \tag{2.1a}$$

$$DV_{xx} + G_{\Phi(s)}(V) = 0 \quad \text{with} \quad V_x = 0 \quad \text{on} \ \partial I \tag{2.1b}$$

where

$$G_{\Phi(s)}(V) = \begin{cases} G_-(V) & \text{on} \ I_- = (0, \varphi(s)) \\ G_+(V) & \text{on} \ I_+ = (\varphi(s), 1). \end{cases} \tag{2.1c}$$

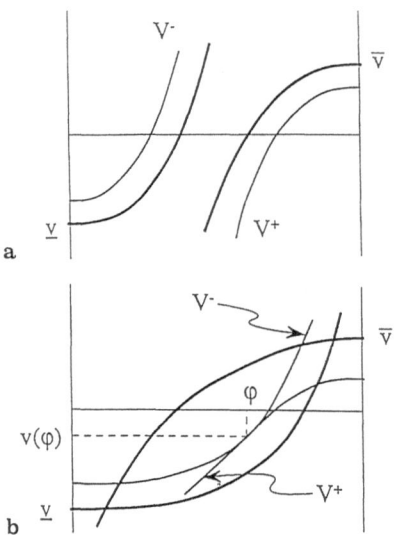

Figure 2.1.

For definiteness we assume that $G_\pm(v)$ take the form as in Figure 1.4 which is equivalent to assume that (f, g) is of Turing type (see Figure 1.1 (b)). More precisely, (A.4) and Remark 1.1 imply that

(G.1) $G_\pm(v)$ are smooth and satisfy $\dfrac{dG_\pm}{dv} < 0$ for $\underline{v} \leq v < \bar{v}$ and $\underline{v} < v \leq \bar{v}$,
   respectively, i.e., strictly monotone decreasing.

(G.2) $G_+(\bar{v}) > 0$ and $G_-(\underline{v}) < 0$.

First we explain the geometrical meaning of C^1-matched solution of (2.1b). We denote a solution of (2.1b) on $I^\pm$ by $V^\pm$, respectively. It is clear from (G.2) that V^- (resp. V^+) is strictly convex (resp. concave) on I^- (resp. I^+) (see Figure 2.1(a)). Also we see from (G.1) and the boundary conditions that V is order-preserving on I^- in the sense that $V_1^-(0) > V_2^-(0)$ implies $V_1^-(x) > V_2^-(x)$ on I^-, and similar property also holds for V^+ on I^+. In view of Figure 2.1(b), it is apparent that, for a given φ, (2.1b) has a C^1-matched solution at $x = \varphi$ if and only if V^+ and V^- are tangent with each other at $x = \varphi$. We denote this solution and its value at $x = \varphi$ by V_φ and $v(\varphi)$, respectively, which are uniquely determined by φ. The slow manifold $\mathcal{M}_1 = \mathcal{M}_1(D)$ for (2.1) is defined by

$$\mathcal{M}_1(D) = \{(\varphi, v(\varphi))|\text{ there is a } C^1\text{ -matched solution to (2.1b) with } 0 < \varphi < 1\}.$$
$$(2.2)$$

In what follows we shall find an analytical expression for $\mathcal{M}_1$ (see (2.7)), which is more convenient to study the flow on it.

It is clear from (G.2) and the order preserving property that $(2.1)_b$ cannot have C^1-matched solutions when D is small (see Figure 2.1(a)), in fact, there is a unique threshold value $D = \underline{D}_1$ such that there are no C^1-matched solutions for $D < \underline{D}_1$. Here $\underline{D}_1$ is defined by

$\underline{D_1}$: When $D = \underline{D}_1$, $(2.1)_b$ has two solutions $V^-(x)$ with $V^-(0) = \underline{v}$ and
$\qquad V^+(x)$ with $V^+(1) = \bar{v}$ which are tangent with each other.

(2.3)

Remark 2.1. *When the nonlinearity (f, g) is of bistable type like Figure 1.1(c), we have no limitation of D for the existence of C^1-matched solutions. In fact, since both G_- and G_+ have zero points, we can always construct C^1-matched solutions of $(2.1)_b$ for any $D(> 0)$ and φ $(0 < \varphi < 1)$.*

To be more precise, it is convenient to introduce the mappings $d^-(\varphi, v)$ and $d^+(1-\varphi, v)$ defined as follows: For a given $\varphi(0 < \varphi < 1)$, consider the boundary value problem

$$DV_{xx} + G_-(V) = 0 \quad \text{on } (0, \varphi) \tag{2.4a}$$

$$V_x(0) = 0, \quad V(\varphi) = v. \tag{2.4b}$$

In view of the assumptions (G.1) and (G.2), it is easy to verify by contradiction that the solution of (2.4), if it exists, is unique. We denote by $d^-(\varphi, v)$ the x-derivative of this solution at $x = \varphi$. Similarly $d^+(1 - \varphi, v)$ is defined as the derivative at $x = 1 - \varphi$ of the solution to

$$DV_{xx} + G_+(V) = 0 \quad \text{on } (0, 1 - \varphi) \tag{2.5a}$$

$$V_x(0) = 0, \quad V(1 - \varphi) = v. \tag{2.5b}$$

Then the C^1-matching condition at $x = \varphi$ can be represented by

$$\mathcal{F}(\varphi, v) = 0, \tag{2.6a}$$

where

$$\mathcal{F}(\varphi, v) \equiv d^-(\varphi, v) + d^+(1 - \varphi, v). \tag{2.6b}$$

Using this expression, we can give an analytical definition for nthe one-dimensional slow manifold in (φ, v)-space:

$$\mathcal{M}_1(D) \equiv \{(\varphi, v) \mid \mathcal{F}(\varphi, v) = 0, 0 < \varphi < 1\}. \tag{2.7}$$

Since

$$d\mathcal{F} = \left(\frac{\partial d^-}{\partial v} + \frac{\partial d^+}{\partial v} \right) dv + \left(\frac{\partial d^-}{\partial \varphi} + \frac{\partial d^+}{\partial \varphi} \right) d\varphi,$$

and from (G.1) and (G.2), we have

$$\frac{\partial d^-}{\partial v} > 0, \quad \frac{\partial d^+}{\partial v} > 0, \quad \frac{\partial d^-}{\partial \varphi} > 0, \quad \frac{\partial d^+}{\partial \varphi} > 0$$

at any solution of (2.6). Let (φ_0, v_0) be any solution of (2.6a), then by using the implicit function theorem, there is a unique smooth curve $v = \bar{v}(\varphi)$ with $v_0 = v(\varphi_0)$ locally near (φ_0, v_0) and $\dfrac{d\bar{v}}{d\varphi} < 0$ holds. $\square$

Using this result and continuous dependency on parameters of solutions, we can prove the following.

Theorem 2.2. *There exists a unique $\underline{D}_1 > 0$ such that*

(1) $\mathcal{M}_1(D) = \phi$ (empty set) for $D < \underline{D}_1$.
(2) For $D > \underline{D}_1$, there exist continuous functions $\underline{\varphi}(D)$ and $\overline{\varphi}(D)$ of $D(0 < \underline{\varphi}(D) < \overline{\varphi}(D) < 1)$ such that $\mathcal{M}_1(D)$ is a smooth one-dimensional manifold defined by

$$\mathcal{M}_1(D) \equiv \{(\varphi, v)|v = \tilde{v}(\varphi) \text{ for } \underline{\varphi}(D) < \varphi < \overline{\varphi}(D)\},$$

i.e., a graph on $(\underline{\varphi}(D), \overline{\varphi}(D))$. Moreover $\dfrac{d\tilde{v}}{d\varphi} < 0$ holds. Finally, suppose a unique equilibrium point $\Phi_1^* = (\varphi^*, v^*)$ is contained in $\mathcal{M}_1(D)$, then it is globally asymptotically stable on $\mathcal{M}_1(D)$.

Proof. (1) is clear from the definition of $\underline{D}_1$ (see (2.3)). As for (2), it follows from the previous discussions that $\mathcal{M}_1(D)$ is locally expressed as a smooth strictly decreasing curve $v = \tilde{v}(\varphi)$. In view of Figure 2.1 and $\dfrac{d\tilde{v}}{d\varphi} < 0$, we see that $\mathcal{M}_1(D)$ is a graph on the interval $(\underline{\varphi}(D), \overline{\varphi}(D))$ where $\overline{\varphi}(D)$ is determined in such a way that V^--solution with $V^-(0) = \underline{v}$ is matched with a V^+-solution at $x = \overline{\varphi}(D)$ (Figure 2.1 (b)), and similaly for $\underline{\varphi}(D)$. Global asymptotic stability of (φ^*, v^*) comes from the fact that $\mathcal{M}_1(D)$ is one-dimensional and (φ^*, v^*) is a unique equilibrium point with $\dfrac{d\tilde{v}}{d\varphi} < 0$ everywhere. □

Remark 2.3. (a) $\mathcal{M}_1(D)$, in general, does not contain the equilibrium point (φ^*, v^*). In fact, the matching value of V^- and V^+ is not, in general, equal to v^* at $D = \underline{D}_1$ (see (2.3)) where $\underline{\varphi}(D)$ and $\overline{\varphi}(D)$ coincide each other, then it holds that $(\varphi^*, v^*) \notin \mathcal{M}_1(D)$ when $D (\geq \underline{D}_1)$ is close to $\underline{D}_1$.

(b) On the other hand, suppose that the nonlinearity (f, g) has an odd symmetry with repect to the middle intersecting point R of $f = 0$ and $g = 0$, the equilibrium point (φ^*, v^*) $(\varphi^* = \dfrac{1}{2})$ always lies in $\mathcal{M}_1(D)$ for $D > \underline{D}_1$.

2.2. Local Slow Manifold for Double-layered Solution

We shall construct a local slow manifold around the equilibrium point for the double-layered case and study the flow on it. In this subsection we fix D to be an appropriate value so that (2.8) has a unique steady state $\Phi_2^* = (\varphi_1^*, \varphi_2^*)$. In view of (1.12), we see that the slow dynamics for two layers can be written as follows:

$$(\varphi_1)_s = c(V(\varphi_1(s))) \tag{2.8a}$$

$$(\varphi_2)_s = -c(V(\varphi_2(s))) \tag{2.8b}$$

$$0 = DV_{xx} + G_{\Phi(s)}(V) \tag{2.8c}$$

$$V_x(0) = V_x(1) = 0 \tag{2.8d}$$

where

$$G_{\Phi(s)}(V) \equiv G_-(V)[H(\varphi_1(s) - x) + H(x - \varphi_2(s))]$$
$$+ G_+(V)H(x - \varphi_1(s))H(\varphi_2(s) - x),$$

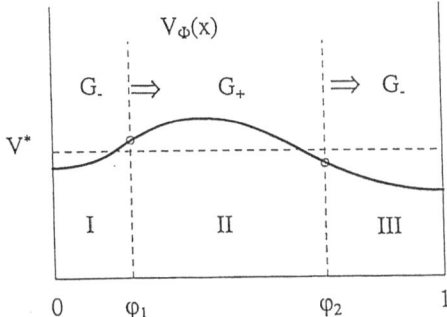

Figure 2.2.

$\Phi(s) = (\varphi_1(s), \varphi_2(s))$, and $H(\xi)$ is the Heaviside function i.e., $H(\xi) = 1$ (resp. 0) for $\xi > 0$ (resp. < 0). Namely $G_\Phi = G_-$ in region I, III, and $G_\Phi = G_+$ in region II (see Figure 2.2). For a given $\Phi = (\varphi_1, \varphi_2)$, we denote the solution of (2.8c) and (2.8d) by $V_\Phi(x)$.

Note that the dynamics (2.8) is valid for $0 < \varphi_1 < \varphi_2 < 1$ and that V_Φ is matched in C^1-sense at layer positions $x = \varphi_i(s)$ $(i = 1, 2)$. Let $(V_\Phi, \varphi_1, \varphi_2) = (V_2, \varphi_1^*, \varphi_2^*)$ be the normal 2-layered solution, i.e., the unique critical point of (2.8). Note that $\varphi_2^* = 1 - \varphi_1^*$ holds because of symmetry. V_2 must satisfy $V_2^*(\varphi_1^*) = v^* = V_2^*(\varphi_2^*)$. What we have to do is to construct the solutions (2.8c), (2.8d) for any (φ_1, φ_2) near $(\varphi_1^*, \varphi_2^*)$ and study the flow on it. Let $(v_1, d_-(\varphi_1, v_1))$ denote the value of V and its derivative at $x = \varphi_1$ by solving (2.8c) in region I. Note that the derivative $d_-(\varphi_1, v_1)$ is uniquely determined as function of (φ_1, v_1) because of the Neumann boundary condition at $x = 0$ and the monotonicity of $G_-(V)$. Using this mapping d_-, we have, for a given v_2, $(v_2, -d_-(1 - \varphi_2, v_2))$ at $x = \varphi_2$. The minus sign in front of d_- is necessary, since we solve (2.8c) from right to left in region III. Finally let $(S_+(v, d, x), d_+(v, d, x))$ denote the solution map of $DV_{xx} + G_+(V) = 0$ for a given initial data $(V, V_x) = (v, d)$ at $x = 0$. Because of symmetry, it is more convenient to make a C^1-matching at $x = \dfrac{\varphi_2 - \varphi_1}{2}$, the middle point of region II:

$$\mathcal{F}_1(\varphi_1, \varphi_2, v_1, v_2) = 0 \tag{2.9a}$$

$$\mathcal{F}_2(\varphi_1, \varphi_2, v_1, v_2) = 0 \tag{2.9b}$$

where

$$\mathcal{F}_1(\varphi_1, \varphi_2, v_1, v_2) \equiv S_+\left(v_1, d_-(\varphi_1, v_1), \frac{\varphi_2 - \varphi_1}{2}\right)$$
$$-S_+\left(v_2, d_-(1 - \varphi_2, v_2), \frac{\varphi_2 - \varphi_1}{2}\right) \tag{2.9c}$$

$$\mathcal{F}_2(\varphi_1, \varphi_2, v_1, v_2) \equiv d_+ \left(v_1, d_-(\varphi_1, v_1), \frac{\varphi_2 - \varphi_1}{2} \right)$$

$$+ d_+ \left(v_2, d_-(1 - \varphi_2, v_2), \frac{\varphi_2 - \varphi_1}{2} \right). \tag{2.9d}$$

At 2-layer solution, it holds that

$$\mathcal{F}_1(\varphi_1^*, \varphi_2^*, v^*, v^*) = 0$$

$$\mathcal{F}_2(\varphi_1^*, \varphi_2^*, v^*, v^*) = 0. \tag{2.10}$$

We show that (2.9) can be solved uniquely with respect to (v_1, v_2) in a neighbourhood of $(\varphi_1^*, \varphi_2^*)$. The mappings $d_\pm$ and S_+ are smooth and the next lemma holds in a neighbourhood of $(v, d, x) = (v^*, d_-(\varphi_1^*, v^*), \varphi_1^*)$ (in fact, it holds in larger region).

Lemma 2.4.

(i)
$$\frac{\partial S_+}{\partial v} > 0, \quad \frac{\partial S_+}{\partial d} > 0$$

(ii)
$$\frac{\partial d_+}{\partial v} > 0, \quad \frac{\partial d_+}{\partial d} > 0, \quad \frac{\partial d_+}{\partial x} < 0$$

(iii)
$$\frac{\partial d_-}{\partial v} > 0, \quad \frac{\partial d_-}{\partial x} > 0$$

Proof. These are the direct consequences of the fact that $G_\pm(V)$ are strictly monotone decreasing and take definite signs, respectively. □

Using Lemma 2.4, we have, at $(v, d, x) = (v^*, d_-(v^*, \varphi_1^*), \varphi_1^*)$,

$$\frac{\partial \mathcal{F}_1}{\partial v_1} = \frac{\partial S_+}{\partial v_1} + \frac{\partial S_+}{\partial d} \frac{\partial d_-}{\partial v_1} > 0$$

$$\frac{\partial \mathcal{F}_1}{\partial v_2} = -\frac{\partial S_+}{\partial v_2} - \frac{\partial S_+}{\partial d} \frac{\partial d_-}{\partial v_2} < 0$$

$$\frac{\partial \mathcal{F}_2}{\partial v_1} = \frac{\partial d_+}{\partial v_1} + \frac{\partial d_+}{\partial d} \frac{\partial d_-}{\partial v_1} > 0$$

$$\frac{\partial \mathcal{F}_2}{\partial v_2} = \frac{\partial d_+}{\partial v_2} + \frac{\partial d_+}{\partial d} \frac{\partial d_-}{\partial v_2} > 0. \tag{2.11}$$

This implies $\det \dfrac{\partial(\mathcal{F}_1, \mathcal{F}_2)}{\partial(v_1, v_2)} > 0$, and hence, by the implicit function theorem, we obtain the following.

Proposition 2.5. *For arbitrary* $\Phi = (\varphi_1, \varphi_2)$ *near* $\Phi_2^* = (\varphi_1^*, \varphi_2^*)$, *there exist a unique solution* $V = V_\Phi(x)$ *of* (2.8)$_c$ *and* (2.8)$_d$ *which converges to normal 2-layer solution in* C^1-*sense when* (φ_1, φ_2) *tends to* $(\varphi_1^*, \varphi_2^*)$. *The values of* $V_\Phi(x)$ *at two layer positions, denoted by* $v_1(\varphi_1, \varphi_2)$, $v_2(\varphi_1, \varphi_2)$, *are uniquely determined and smooth functions of* (φ_1, φ_2) *in a neighbourhood of* $(\varphi_1^*, \varphi_2^*)$.

Using Proposition 2.5, (2.8a) and (2.8b), the vector field on a local slow manifold is given by

$$\frac{d}{ds}\varphi_1 = c(v_1(\varphi_1, \varphi_2))$$

$$\frac{d}{ds}\varphi_2 = -c(v_2(\varphi_1, \varphi_2)).$$

(2.12)

Hence *the original dynamical system* (2.8) *is contracted to two-dimensional nonlinear ODE system* (2.12).

Now it is straightforward to check the stability of critical point $\Phi_2^* = (\varphi_1^*, \varphi_2^*)$ by computing the linearized matrix of (2.12) at Φ_2^*. Namely it suffices to find the signs of the real parts of eigenvalues of

$$J_{\Phi_2^*} \equiv \left(\begin{array}{cc} \dfrac{\partial c}{\partial v_1}\dfrac{\partial v_1}{\partial \varphi_1} & \dfrac{\partial c}{\partial v_1}\dfrac{\partial v_1}{\partial \varphi_2} \\[2ex] -\dfrac{\partial c}{\partial v_2}\dfrac{\partial v_2}{\partial \varphi_1} & -\dfrac{\partial c}{\partial v_2}\dfrac{\partial v_2}{\partial \varphi_2} \end{array} \right) \Bigg|_{\Phi_2^*}$$

(2.13)

We shall show $\det J_{\Phi_2^*} > 0$ and $\operatorname{tr} J_{\Phi_2^*} < 0$. From the differential relations,

$$\frac{\partial \mathcal{F}_1}{\partial v_1}dv_1 + \frac{\partial \mathcal{F}_1}{\partial v_2}dv_2 + \frac{\partial \mathcal{F}_1}{\partial \varphi_1}d\varphi_1 + \frac{\partial \mathcal{F}_1}{\partial \varphi_2}d\varphi_2 = 0$$

$$\frac{\partial \mathcal{F}_2}{\partial v_1}dv_1 + \frac{\partial \mathcal{F}_2}{\partial v_2}dv_2 + \frac{\partial \mathcal{F}_2}{\partial \varphi_1}d\varphi_1 + \frac{\partial \mathcal{F}_2}{\partial \varphi_2}d\varphi_2 = 0$$

we have

$$\left(\begin{array}{c} dv_1 \\[2ex] dv_2 \end{array} \right) = \frac{1}{\dfrac{\partial(\mathcal{F}_1, \mathcal{F}_2)}{\partial(v_1, v_2)}} \left(\begin{array}{cc} \dfrac{\partial \mathcal{F}_2}{\partial v_2} & -\dfrac{\partial \mathcal{F}_1}{\partial v_2} \\[2ex] -\dfrac{\partial \mathcal{F}_2}{\partial v_1} & \dfrac{\partial \mathcal{F}_1}{\partial v_1} \end{array} \right) \left(\begin{array}{cc} -\dfrac{\partial \mathcal{F}_1}{\partial \varphi_1} & -\dfrac{\partial \mathcal{F}_1}{\partial \varphi_2} \\[2ex] -\dfrac{\partial \mathcal{F}_2}{\partial \varphi_1} & -\dfrac{\partial \mathcal{F}_2}{\partial \varphi_2} \end{array} \right) \left(\begin{array}{c} d\varphi_1 \\[2ex] d\varphi_2 \end{array} \right).$$

(2.14)

It is obvious that (i, j) -component of the matrix on the right-hand side of (2.14) is equal to $\dfrac{\partial v_i}{\partial \varphi_j}$, i.e.,

$$\left\{ \frac{\partial v_i}{\partial \varphi_j} \right\}_{i,j=1,2} \equiv \left\{ \frac{\partial(\mathcal{F}_1, \mathcal{F}_2)}{\partial(v_1, v_2)} \right\}^{-1} \left\{ -\frac{\partial \mathcal{F}_i}{\partial \varphi_j} \right\}_{1\leq i,j\leq 2}$$

(2.15)

By similar computation to (2.11), we easily see that

$$\frac{\partial \mathcal{F}_1}{\partial \varphi_1} > 0, \quad \frac{\partial \mathcal{F}_1}{\partial \varphi_2} > 0, \quad \frac{\partial \mathcal{F}_2}{\partial \varphi_1} > 0, \text{ and } \quad \frac{\partial \mathcal{F}_2}{\partial \varphi_2} < 0,$$

which implies $\dfrac{\partial(\mathcal{F}_1, \mathcal{F}_2)}{\partial(v_1, v_2)} < 0$, and hence, $\det \left\{ \dfrac{\partial v_i}{\partial \varphi_j} \right\}_{i,j=1,2} < 0$. Noting that $\dfrac{\partial c}{\partial v_1} =$

$\dfrac{\partial c}{\partial v_2} > 0$ at Φ_2^*, we obtain

$$\det J_{\Phi_2^*} = - \left(\left. \dfrac{\partial c}{\partial v_1} \right|_{\Phi_2^*} \right)^2 \det \left\{ \dfrac{\partial v_i}{\partial \varphi_j} \right\}_{i,j=1,2} > 0.$$

On the other hand, we have

$$\operatorname{tr} J_{\Phi_2^*} = \left. \dfrac{\partial c}{\partial v_1} \right|_{\Phi_2^*} \left. \left\{ \dfrac{\partial v_1}{\partial \varphi_1} - \dfrac{\partial v_2}{\partial \varphi_2} \right\} \right|_{\Phi_2^*}.$$

In view of (2.11), (2.14), (2.15), and (2.16), we easily see that $\operatorname{tr} J_{\Phi_2^*} < 0$. Thus we conclude that

Proposition 2.6. *The critical point* $(\varphi_1, \varphi_2) = (\varphi_1^*, \varphi_2^*)$ *corresponding to normal 2-layer solution is asymptotically stable equilibrium of* (2.8) *on the local slow manifold constructed in Proposition 2.5.*

In the rest of this subsection we consider the stability of Φ_2^* from more intuitive point of view. Essentially there are two types of perturbation; symmetric and anti-symmetric ones as in Figure 2.3.

The first case (Figure 2.3(a)) can be reduced to the mono-layer case. Namely, it suffices to see only the half part because of symmetry, and hence it was already studied in Section 2.1. The second case (Figure 2.3(b)), the anti-symmetric perturbation, where the directions of the shifts of two layer positions are the same, is the main concern here. Recalling that the velocity function $c(V)$ is monotone decreasing and $c(V(\varphi_i^*)) = 0$ ($i = 1, 2$), we see that, in order to stabilize this perturbation, the v-value at left (resp. right) layer must be up (resp. down) so that they recover their original positions. We will see that this really occurs as in Figure 2.3(b). Arrows in Figure 2.3 indicate the directions in which the perturbed layers move. Since the distance between φ_1 and φ_2 does not change for the anti-symmetric perturbation, it holds from $\varphi_2^* = 1 - \varphi_1^*$ that

$$\varphi_2 - \varphi_1 = 1 - 2\varphi_1^*. \qquad (2.17)$$

In order to know how v_1 and v_2 behave on the slow manifold, we consider the differentials of (2.9):

$$d\mathcal{F}_1 = \dfrac{\partial \mathcal{F}_1}{\partial v_1} dv_1 + \dfrac{\partial \mathcal{F}_1}{\partial v_2} dv_2 + \dfrac{\partial \mathcal{F}_1}{\partial \varphi_1} d\varphi_1 + \dfrac{\partial \mathcal{F}_1}{\partial \varphi_2} d\varphi_2 = 0$$

$$d\mathcal{F}_2 = \dfrac{\partial \mathcal{F}_2}{\partial v_1} dv_1 + \dfrac{\partial \mathcal{F}_2}{\partial v_2} dv_2 + \dfrac{\partial \mathcal{F}_2}{\partial \varphi_1} d\varphi_1 + \dfrac{\partial \mathcal{F}_2}{\partial \varphi_2} d\varphi_2 = 0 \qquad (2.18)$$

We compute the directional derivative of v_1 and v_2 along the line (2.17). Recalling the form (2.9), we see that

$$\dfrac{\partial \mathcal{F}_1}{\partial \varphi_1} = \dfrac{\partial \mathcal{F}_1}{\partial \varphi_2} \quad \text{and} \quad \dfrac{\partial \mathcal{F}_2}{\partial \varphi_1} + \dfrac{\partial \mathcal{F}_2}{\partial \varphi_2} = 0 \qquad (2.19)$$

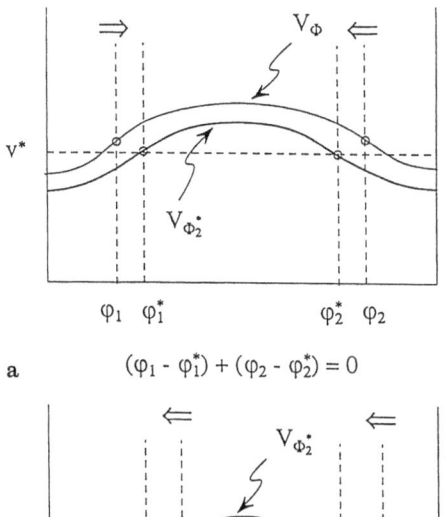

a $(\varphi_1 - \varphi_1^*) + (\varphi_2 - \varphi_2^*) = 0$

b $(\varphi_1 - \varphi_1^*) - (\varphi_2 - \varphi_2^*) = 0$

Figure 2.3.

hold at $(\varphi_1, \varphi_2, v_1, v_2) = (\varphi_1^*, \varphi_2^*, v^*, v^*)$. Also, by a simple computation, we see from Lemma 2.4 that $\dfrac{\partial \mathcal{F}_1}{\partial v_1} > 0$, $\dfrac{\partial \mathcal{F}_1}{\partial v_2} < 0$, $\dfrac{\partial \mathcal{F}_1}{\partial \varphi_1} > 0$, $\dfrac{\partial \mathcal{F}_2}{\partial v_1} > 0$, $\dfrac{\partial \mathcal{F}_2}{\partial v_2} > 0$, holds. This combined with (2.18) and (2.19) implies that

$$D^+ v_1 < 0 \text{ and } D^+ v_2 > 0 \text{ at } (\varphi_1^*, \varphi_2^*) \tag{2.20}$$

where D^+ denotes the directional derivative along the line (2.17), i.e., (1,1)-direction. This is what we expected before.

2.3. Formal Linearization

In this subsection we shall derive a formal linearized equation for the singular limit equations (2.8). The reason for this is that the resulting linearized eigenvalue problem is exactly the same form as the SLEP system which will be derived from the original system in a rigorous manner in Section 3. This is important since formal analysis gives us a correct answer for the original system. As was mentioned in Section 1, the dynamics has two steps; first, the outer dynamics, and then slow layer dynamics. So is the linearized problem. We will show that the linearized problem for outer dynamics is quite stable, i.e., all the spectrum have strictly negative parts, on the other hand, the one for the layer dynamics, which corresponds to the SLEP system, is subtle and needs more computation.

A. Outer Part

The dynamics in this region is given by

$$\delta U_t = f(U, V)$$

$$V_t = \frac{1}{\sigma} V_{xx} + g(U, V),$$

(2.21)

where we replace D by $1/\sigma$ for later convenience. Let (U_2^*, V_2^*) be the normal 2-layer solution associated with Φ_2^*, i.e., V_2^* is the solution of (2.8c) and (2.8d) with $\Phi = \Phi_2^*$, and U_2^* is the associated u -component through the relation $u = h_\pm(v)$. In Section 3 we will write this solution with superscript σ as in Corollary 3.9, however we omit it here for simplicity. We set $U = U_2^* + e^{\lambda t} W$, $V = V_2^* + e^{\lambda t} Z$ where (W, Z) is a perturbation in outer region which become zero at layer positions. Substituting this into (2.21), the resulting linearized problem becomes

$$\delta \lambda W = f_u^* W + f_v^* Z$$

$$\lambda Z = \frac{1}{\sigma} Z_{xx} + g_u^* W + g_v^* Z,$$

(2.22)

subject to $Z_x = 0$ at $x = 0, 1$, where $f_u^* \equiv f_u(U_2^*, V_2^*)$ and so on. Since $f_u^* < 0$ (see (A.3)) we can solve the first equation of (2.22) as $W = -f_v^* Z/(f_u^* - \delta \lambda)$ which is well-defined as far as λ varies in

$$C_\mu = \{\lambda \mid Re\delta\lambda > -\mu, \quad 0 > -\mu > \inf_{x \neq \varphi_1^*, \varphi_2^*} f_u(U_2^*(x), V_2^*(x))\}.$$

Clearly this restriction does not affect our stability analysis. Substituting this into the second one of (2.22), we obtain

$$\lambda Z = \frac{1}{\sigma} Z_{xx} + \left(-\frac{f_v^* g_u^*}{f_u^* - \delta \lambda} + g_v^* \right) Z.$$

(2.23)

It is known that this problem exactly coincides with the eigenvalue problem for the *noncritical* eigenvalues (see Section 2 of [44]). Typically, $f_v^* < 0$, $g_u^* > 0$, and $g_v^* < 0$ (for instance, $f(u, v) = u(1 - u)(u - a) - v$ and $g(u, v) = \gamma u - \sigma v$ with $\gamma > 0$, $\sigma > 0$), hence the coefficient of Z on the right-hand side of (2.23) is strictly negative. Making a bilinear form, we easily see that all the spectrum of (2.23) have strictly negative real parts. Thus there are no dangerous eigenvalues coming from the outer part.

B. Layer Part

Using the *slow* time $s = \varepsilon t/\delta$, the layer dynamics was given by (2.8). $(V, \varphi_1, \varphi_2) = (V_2^*(x), \varphi_1^*, \varphi_2^*)$ is the normal 2-layer solution. Recall that $V_2^*(\varphi_1^*) = V_2^*(\varphi_2^*) = v^*$ and

$c(v^*) = 0$. We derive the linearized equations at $(V_2^*, \varphi_1^*, \varphi_2^*)$ by setting

$$V = V_2^*(x) + e^{\tau s}h(x)$$

$$\varphi_1 = \varphi_1^* - e^{\tau s}\psi_1 \qquad (2.24)$$

$$\varphi_2 = \varphi_2^* - e^{\tau s}\psi_2$$

Here we employ $\tau \equiv \lambda\delta/\varepsilon$ as the eigenvalue parameter, since we adopt the slow time s instead of t. Note that we put the minus sign in front of $e^{\lambda s}\psi_1$ in order for ψ_1 and ψ_2 to have the same sign for anti-symmetric perturbation (see Figure 2.3(b)). Substituting (2.24) into (2.8), and linearizing it in a formal way, we have

$$-\tau\psi_1 = \frac{dc}{dV}(v^*)\left\{-\frac{dV_2^*}{dx}(\varphi_1^*)\psi_1 + h(\varphi_1^*)\right\} \qquad (2.25a)$$

$$\tau\psi_2 = -\frac{dc}{dV}(v^*)\left\{\frac{dV_2^*}{dx}(\varphi_2^*)\psi_2 + h(\varphi_2^*)\right\}, \qquad (2.25b)$$

$$0 = \frac{1}{\sigma}h_{xx} + \frac{dG_-}{dV}(V_2^*)\{H(\varphi_1^* - x) + H(x - \varphi_2^*)\}h \qquad (2.25c)$$

$$+\frac{dG_+}{dV}(V_2^*)\{H(x - \varphi_1^*)H(\varphi_2^* - x)\}h$$

$$+G_-(V_2^*)(-\delta_{\varphi_1^*}\psi_1 - \delta_{\varphi_2^*}\psi_2)$$

$$+G_+(V_2^*)(\delta_{\varphi_1^*}\psi_1)H(\varphi_2^* - x)$$

$$+G_+(V_2^*)H(x - \varphi_1^*)\delta_{\varphi_2^*}\psi_2$$

where $\delta_{\varphi_i^*}$ is the Dirac's δ -function with support at $x = \varphi_i^*$. Using

$$H(\varphi_2^* - x)\delta_{\varphi_1^*}\psi_1 = \delta_{\varphi_1^*}\psi_1, \qquad H(x - \varphi_1^*)\delta_{\varphi_2^*}\psi_2 = \delta_{\varphi_2^*}\psi_2,$$

and

$$\{G_+(V_2^*) - G_-(V_2^*)\}\delta_{\varphi_i^*}\psi_i = [g]\delta_{\varphi_i^*}\psi_i \qquad (i = 1, 2),$$

where $[g] = g(h_+(v^*), v^*) - g(h_-(v^*), v^*)$, i.e., the jump of the value of g from left- to right-branch at $v = v^*$, (2.25c) becomes

$$T^{*,\sigma}h = [g]\left(\delta_{\varphi_1^*}\psi_1 + \delta_{\varphi_2^*}\psi_2\right), \qquad (2.26)$$

where

$$T^{*,\sigma} \equiv -\frac{1}{\sigma}\frac{d^2}{dx^2} - \frac{dG_-}{dV}(V_2^*)\{H(\varphi_1^* - x) + H(x - \varphi_2^*)\}$$

$$-\frac{dG_+}{dV}(V_2^*)H(x - \varphi_1^*)H(\varphi_2^* - x).$$

Since $G_\pm(s)$ is strictly monotone decreasing (see (G.1)), i.e., $\dfrac{dG_\pm}{dV} < 0$, $T^{*,\sigma}$ has a well-defined inverse $K^{*,\sigma,0} : H^{-1}(I) \to H_N^1(I)$. Hence it follows from (2.26) that

$$h = [g]K^{*,\sigma,0}\left(\delta_{\varphi_1^*}\psi_1 + \delta_{\varphi_2^*}\psi_2\right). \tag{2.27}$$

This leads to the expression:

$$h(\varphi_1^*) = [g]\langle K^{*,\sigma,0}\left(\delta_{\varphi_1^*}\psi_1 + \delta_{\varphi_2^*}\psi_2\right), \delta_{\varphi_1^*}\rangle$$

$$h(\varphi_2^*) = [g]\langle K^{*,\sigma,0}\left(\delta_{\varphi_1^*}\psi_1 + \delta_{\varphi_2^*}\psi_2\right), \delta_{\varphi_2^*}\rangle$$

Substituting these into (2.25a), (2.25b), we have

$$-\tau\psi_1 = \frac{dc}{dV}(v^*)\left\{-\frac{dV_2^*}{dx}(\varphi_1^*)\psi_1 + [g]\langle K^{*,\sigma,0}\left(\delta_{\varphi_1^*}\psi_1 + \delta_{\varphi_2^*}\psi_2\right), \delta_{\varphi_1^*}\rangle\right\}$$

$$\tau\psi_2 = \frac{dc}{dV}(v^*)\left\{\frac{dV_2^*}{dx}(\varphi_2^*)\psi_2 + [g]\langle K^{*,\sigma,0}\left(\delta_{\varphi_1^*}\psi_1 + \delta_{\varphi_2^*}\psi_2\right), \delta_{\varphi_2^*}\rangle\right\} \tag{2.28}$$

In order to compare (2.28) with the SLEP system (3.75) in Section 3, we need the following equalities. For the definitions of notation, see Remark 3.7, Lemmas 3.15 and 3.22 in Section 3.

Lemma 2.7.

$$\frac{dc}{dV}(v^*)\frac{dV_2^*}{dx}(\varphi_1^*) = -\frac{dc}{dV}(v^*)\frac{dV_2^*}{dx}(\varphi_2^*) = \sigma\zeta^{*,\sigma} \tag{i}$$

$$-\frac{dc}{dV}(v^*)[g] = \frac{1}{\left\|\dfrac{d}{dy}\tilde{u}^*\right\|_{L^2}^2}\frac{dJ}{dv}(v^*)[g] \tag{ii}$$

$$= -\frac{1}{\left\|\gamma^*\dfrac{d}{dy}\tilde{u}^*\right\|_{L^2}^2}\left(-\gamma^*\frac{dJ}{dv}(v^*)\right)(\gamma^*[g])$$

$$= -\frac{c_1^* c_2^*}{(c^*)^2}.$$

Proof. See Appendix A. □

Using Lemma 2.7, we can rewrite (2.28) in an equivalent form:

$$\left[\frac{c_1^* c_2^*}{(c^*)^2} \left(\begin{array}{cc} \langle K^{*,\sigma,0}\delta_{\varphi_1^*}, \delta_{\varphi_1^*} \rangle & \langle K^{*,\sigma,0}\delta_{\varphi_2^*}, \delta_{\varphi_1^*} \rangle \\ \langle K^{*,\sigma,0}\delta_{\varphi_1^*}, \delta_{\varphi_2^*} \rangle & \langle K^{*,\sigma,0}\delta_{\varphi_2^*}, \delta_{\varphi_2^*} \rangle \end{array} \right) + (\tau - \sigma\hat{\zeta}^{*,\sigma})I \right] \left(\begin{array}{c} \psi_1 \\ \psi_2 \end{array} \right) = 0. \quad (2.29)$$

This is exactly the *same* as the SLEP system (see (3.75)) for 2- layered case. According to Theorem 3.24, (2.29) has two negative real eigenvalues, which implies the stability of normal 2-layer solutions. In a similar way, but tedious, we can derive the same result for the n-layered case by linearizing the singular limit slow equations (1.12). We leave the details to the reader.

Remark 2.8. *In view of the above computation, the form (2.29) does not depend on the behavior of δ as far as it belongs to the regime $\varepsilon/\delta = o(1)$ as $\varepsilon \downarrow 0$, although the asymptotic form of the resulting critical eigenvalues depends on δ (i.e., it behaves like $O(\varepsilon/\delta)$). It should be noted that, when $\delta = o(1)$ as $\varepsilon \downarrow 0$, u reacts much faster than v and hence the system belongs to the third category according to the classification in Section 1. This suggests that more careful classification is needed depending on the asymptotic regime of δ and ε. In fact, as we saw, as far as stability properties of steady states are concerned, there are no drastic change in the regime of $\varepsilon/\delta = o(1)$ as $\varepsilon \downarrow 0$, however when δ becomes comparable to ε, the system behaves in a quite different manner (see Section 5).*

Remark 2.9. *Although, for 2-layer case, we treat only the local dynamics near Φ_2^*, the global picture has been clarified recently for the bistable case (see Figure 1.1(c)) by [48]: (φ_1, φ_2)-space is divided into three regions by two separation orbits and Φ_2^* lies in one of the domains, say Ω_2^*. If the initial data belongs to Ω_2^*, the orbit tends to Φ_2^* as $t \uparrow \infty$. If not, one of the layers hit the boundary and disappear at certain finite time, then it approaches a mono-layered solution.*

3. The SLEP Method for the Stability of Normal N-layered Solutions

In the previous sections we derived the singular limit slow dynamics (1.12) which approximates the dynamics of the original system (1.1) after layers are fully developed, and proved that all the equilibrium points $\{\Phi_n^*\}_{n=1}$ are locally asymptotically stable. However we do not know in a rigorous sense that how the total dynamics of (1.12) is close to that of (1.1). We shall show in this section by means of the SLEP method that, as far as local stability is concerned, the formal linearized stability analysis for (1.12), which was done in Section 2, becomes a true criterion for the original system (1.1). In fact (1.1) has an ε-family of the normal n-layered solutions (Corollary 3.9) which tends to Φ_n^* as $\varepsilon \downarrow 0$, and they all become stable for sufficiently small ε. This immediately leads us to Main Theorem and the commutative diagram (Figure 1.8) in Section 1. Note that linearized stability implies nonlinear stability for the semilinear parabolic system (1.1) (see, for instance, Henry [29]). This section is a detailed version of the paper Nishiura and Fujii [45].

In what follows, we use the notation $1/\sigma$ instead of D for the diffusion coefficient of v, and assume that $\delta = 1$, since, for general δ, a similar result can be obtained by obvious modification (see Remark 2.8). The model system and its stationary problem become

$$
\begin{cases}
u_t &= \varepsilon^2 u_{xx} + f(u, v) \\[2mm]
v_t &= \dfrac{1}{\sigma} v_{xx} + g(u, v)
\end{cases}
\tag{3.1}
$$

and

$$
\begin{cases}
0 &= \varepsilon^2 u_{xx} + f(u, v) \\[2mm]
0 &= \dfrac{1}{\sigma} v_{xx} + g(u, v),
\end{cases}
\tag{3.2}
$$

respectively. The boundary conditions are always Neumann ones unless otherwise stated.

The core of the SLEP method lies in the asymptotic characterizations of critical eigenvalues and the associated eigenfunctions which are valid up to $\varepsilon = 0$. To do this, the study of the spectral behavior of the singular Sturm-Liouville operator $L^{\varepsilon, \sigma}$ (see (3.12b)) is basic. We start this section with the definition of the normal n-layered solution (Figure 1.3).

3.1. Normal N-layered Solution

In this subsection we briefly mention about the construction of, what we call, the normal n-layered solution to (3.2) which converge to the equilibrium point Φ_n^* of (1.12) as $\varepsilon \downarrow 0$. This solution consists of two parts; the *outer* and *inner* parts. The outer one is determined by the formal limiting system of (3.2) (see (3.3)), and the inner one is essentially, after stretching, the stationary front of (1.8) with $V(\varphi) = v^*$ which compensates the jump discontinuity of u from h_--branch to h_+-branch. Here we simply collect necessary results for later discussions. For the detailed proofs of them, see Fife [16], Mimura, Tabata, and Hosono [39], Ito [32], and Nishiura and Fujii [44; appendix].

First we construct a mono-layered solution, which we call the *basic pattern*, then apply the folding up principle (Proposition 3.8) to it to obtain the normal n-layered solutions. The solution of the following reduced problem becomes the first approximation to the basic pattern in outer region.

$$
f(u, v) = 0,
\tag{3.3a}
$$

$$
(u, v) \in L^2(I) \times \{H^2(I) \cap H_N^1(I)\},
$$

$$
\frac{1}{\sigma} v_{xx} + g(u, v) = 0.
\tag{3.3b}
$$

Since we are interested in the solutions of (3.3) which are the limit of those of (3.2) as $\varepsilon \downarrow 0$, we take

$$
u = h^*(v) \equiv
\begin{cases}
h_-(v) & \text{for } v \leq v^*, \\[2mm]
h_+(v) & \text{for } v \geq v^*
\end{cases}
\tag{3.4}
$$

as a special solution of (3.3a), i.e., u has a jump from h_--branch to h_+-branch at $v = v^*$. The reason why we take this special value $v = v^*$ comes from the fact that the velocity of the inner front must be zero in order to have a stationary solution (see (1.8), (1.9), and (3.10)). Substituting this into (3.3b), we obtain the reduced scalar equation for v:

$$\frac{1}{\sigma} v_{xx} + G^*(v) = 0, \qquad v \in H^2(I) \cap H^1_N(I), \tag{3.5}$$

where $G^*(v) \equiv g(h^*(v), v)$. Since $G^*(v)$ has a jump discontinuity at $v = v^*$ (see Figure 1.4), the solution of (3.5) has to be matched in C^1-sense at this switching value.

Lemma 3.1. *There exists a uniquely determined positive constant σ_1^* such that monotone increasing (resp. decreasing) C^1-matched solution $V_+^{*,\sigma}(x)$ (resp. $V_-^{*,\sigma}(x)$) of (3.5) exists uniquely for $0 < \sigma \leq \sigma_1^*$. Moreover, we have $\lim_{\sigma \downarrow 0} V^{*,\sigma}(x) = v^*$ in C^1-sense.*

For definiteness, we only consider the monotone increasing case and write simply $V^{*,\sigma}(x)$ instead of $V_+^{*,\sigma}(x)$. In view of (3.4), the first approximate solution takes the following form:

$$(U^{*,\sigma}(x), V^{*,\sigma}(x)) \qquad \text{for } 0 \leq \sigma \leq \sigma_1^*, \tag{3.6}$$

where $U^{*,\sigma}(x) \equiv h^*(V^{*,\sigma}(x))$. We call (3.6) the *reduced solution* for the basic pattern.

Corollary 3.2. *The matching point $x_1^*(\sigma)$ is well-defined by*

$$V^{*,\sigma}(x_1^*(\sigma)) = v^* \tag{3.7}$$

due to the monotonicity of $V^{*,\sigma}(x)$. Then $x_1^*(\sigma)$ becomes a continuous function for $0 \leq \sigma \leq \sigma_1^*$.

Applying the singular perturbation techniques to (3.6), we have the following existence result for the basic pattern ([16], [39], [32], [44]).

Theorem 3.3 (Existence Theorem for the Basic Pattern). *For any σ_0 with $0 < \sigma_0 < \sigma_1^*$, there is an $\varepsilon_0 > 0$ such that (3.2) has an (ε, σ)-family of solutions $D^1(\varepsilon, \sigma) = (u^1(x; \varepsilon, \sigma), v^1(x; \varepsilon, \sigma)) \in C_\varepsilon^2(\bar{I}) \times C^2(\bar{I})$ for $(\varepsilon, \sigma) \in Q^1 = \{(\varepsilon, \sigma)|0 < \varepsilon < \varepsilon_0, 0 \leq \sigma \leq \sigma_0\}$. $D^1(\varepsilon, \sigma)$ are uniformly bounded in $C_\varepsilon^2(\bar{I}) \times C^2(\bar{I})$, and satisfy*

$$\lim_{\varepsilon \downarrow 0} u^1(x; \varepsilon, \sigma) = U^{*,\sigma}(x) \text{ uniformly on } \bar{I} \backslash I_\kappa \text{ for any } \kappa > 0 \tag{3.8a}$$

and

$$\lim_{\varepsilon \downarrow 0} v^1(x; \varepsilon, \sigma) = V^{*,\sigma}(x) \text{ uniformly on } \bar{I} \tag{3.8b}$$

where $I_\kappa = (x_1^*(\sigma) - \kappa, x_1^*(\sigma) + \kappa)$. Moreover, $D^1(\varepsilon, \sigma)$ depends continuously on $(\varepsilon, \sigma) \in Q^1$ in $C_\varepsilon^2 \times C^2$-topology, and continuously on $(\varepsilon, \sigma) \in \bar{Q}^1$ in $L^2 \times C^1$-topology.
Using a streched variable y defined by

$$y \equiv \frac{x - x_1^*(\sigma)}{\varepsilon}, \qquad y \in \tilde{I} \equiv \left(-\frac{x_1^*(\sigma)}{\varepsilon}, \frac{1 - x_1^*(\sigma)}{\varepsilon} \right)$$

the inner part of $D_1(\varepsilon, \sigma)$ behaves as

$$\lim_{\varepsilon \downarrow 0} (\tilde{u}^{\varepsilon,\sigma}(y), \tilde{v}^{\varepsilon,\sigma}(y)) = (\tilde{u}^*(y), v^*) \quad in \;\; C^2_{c.u.}(\mathbf{R})\text{-sense,} \tag{3.9}$$

where $(\tilde{u}^{\varepsilon,\sigma}, \tilde{v}^{\varepsilon,\sigma})$ are the stretched solutions defined by

$$\tilde{u}^{\varepsilon,\sigma}(y) \equiv u^1(x_1^* + \varepsilon y; \varepsilon, \sigma), \qquad \tilde{v}^{\varepsilon,\sigma}(y) \equiv v^1(x_1^* + \varepsilon y; \varepsilon, \sigma).$$

$\tilde{u}^*(y)$ is a translate of the unique monotone increasing solution of

$$\begin{cases} \dfrac{d^2}{dy^2}\tilde{u} + f(\tilde{u}, v^*) = 0 \\[2mm] \tilde{u}(\pm\infty) = h_\pm(v^*) \\[2mm] \tilde{u}(0) = h_0(v^*), \end{cases} \tag{3.10}$$

and v^* is the unique zero of $J(v)$ (see (A.2)). Here we use the convention of Remark 3.5 for the stretched functions. Note that the limiting function of (3.9) does not depend on σ.

Remark 3.4. The convergence result for $\tilde{v}^{\varepsilon,\sigma}$ (see (3.9)) comes from the following fact: Let $\varphi^\varepsilon(x) \in C^0(\bar{I})$ for $0 < \varepsilon < \varepsilon_0$, and $\|\varphi^\varepsilon\|_{C^2(I)}$ be uniformly bounded with respect to ε. Suppose that $\lim_{\varepsilon \downarrow 0} \varphi^\varepsilon(x^*) = \alpha$ holds at some point $x^* \in I$, where α is a constant. Then the stretched function $\tilde{\varphi}^\varepsilon(y)$ at $x = x^*$, i.e., $\tilde{\varphi}^\varepsilon(y) \equiv \varphi^\varepsilon(x^* + \varepsilon y)$ satisfies

$$\lim_{\varepsilon \downarrow 0} \tilde{\varphi}^\varepsilon(y) = \alpha \quad in \;\; C^2_{c.u.}(\mathbf{R})\text{-sense.}$$

Remark 3.5. For a given stretched function $\tilde{\varphi}(y)(y \in \bar{I})$, it is convenient to extend the definition domain from $\bar{I}$ to the whole line $\mathbf{R}$ in a smooth way so that $\varphi \equiv 0$ for large $|y|$. We use the same notation for the extended one.

Corollary 3.6. Let $F(u, v)$ be a smooth function of u and v. Then, the composite function $F(\tilde{u}^{\varepsilon,\sigma}, \tilde{v}^{\varepsilon,\sigma})$ converges to $F(\tilde{u}^*, v^*)$ in $C^2_{c.u.}(\mathbf{R})$-sense.

Remark 3.7. Differentiating (3.10) by y, we see that $\dfrac{d}{dy}\tilde{u}^*(> 0)$ is a constant multiple of the principal eigenfunction of the following eigenvalue problem associated with the principal eigenvalue $\zeta = 0$:

$$\hat{L}^*\tilde{\phi} \equiv \frac{d^2}{dy^2}\tilde{\phi} + f_u(\tilde{u}^*, v^*)\tilde{\phi} = \zeta\tilde{\phi} \qquad on \;\; \mathbf{R}, \qquad \tilde{\phi} \in L^2(\mathbf{R}). \tag{3.11}$$

We denote by $\hat{\phi}_0^*$ and $\tilde{\phi}_L^*$ the L^2-normalized principal eigenfunction and the positive L^1-normalized principal eigenfunction, respectively, i.e., $\| \hat{\phi}_0^* \|_{L^2(\mathbf{R})} = 1$, $\displaystyle\int_{\mathbf{R}} \tilde{\phi}_L^* dx = 1$. The interrelation among these quantities is given by

$$\hat{\phi}_0^* = \frac{\gamma^*}{c^*} \frac{d}{dy}\tilde{u}^*$$

$$\hat{\phi}_L^* = \gamma^* \frac{d}{dy} \tilde{u}^*$$

$$\hat{\phi}_L^* = c^* \hat{\phi}_0^*,$$

where

$$c^* \equiv \|\hat{\phi}_L^*\|_{L^2(\mathbf{R})}$$

$$\gamma^* \equiv 1/(h_+(v^*) - h_-(v^*)).$$

Multi-layered patterns called the normal n-layered solutions are easily constructed by applying the next proposition to the basic pattern.

Proposition 3.8 (Folding Up Principle). *Suppose $W(x; \underline{d})$ is a solution of (3.2) at $\underline{d} = (\tilde{\varepsilon}^2, \tilde{\sigma}^{-1})$, then $R^n(W)(x)$ is a solution of (3.2) at $\underline{d}/n^2 = (\varepsilon^2, \sigma^{-1})$ for $n = 1, 2, \cdots$, where $(\varepsilon, \sigma) \equiv (\tilde{\varepsilon}/n, n^2\tilde{\sigma})$. Here*

$$R^n(W)(x) = \begin{cases} W(n(x - i/n); \underline{d}) & i = even, \\ \\ W(n(1/n - (x - i)/n); \underline{d}) & i = odd, \end{cases}$$

for $i/n \le x \le (i+1)/n$ ($i = 0, 1, 2, \cdots, n-1$).

Intuitively speaking, $R^n(W)$ can be obtained by flipping W n-times and normalizing the length of interval to one.

Corollary 3.9 (Existence of the Normal n-layered solution). *Let $W = D^1(\tilde{\varepsilon}, \tilde{\sigma})((\tilde{\varepsilon}, \tilde{\sigma}) \in Q^1)$ in Proposition 3.8, then $D^n(\varepsilon, \sigma) = (u^n(x; \varepsilon, \sigma), v^n(x; \varepsilon, \sigma))$ defined by $R^n(D^1(\tilde{\varepsilon}, \tilde{\sigma}))$ becomes a solution of (3.2) with n interior transition layers for $(\varepsilon, \sigma) = (\tilde{\varepsilon}/n, n^2\tilde{\sigma}) \in Q^n \equiv \{(\varepsilon, \sigma) | 0 < \varepsilon < \varepsilon_0/n, 0 \le n^2\sigma_0\}$ (see Figure 1.3). We write the reduced solution of $D^n(\varepsilon, \sigma)$ (i.e., the L^2-limit of $D^n(\varepsilon, \sigma)$ as $\varepsilon \downarrow 0$) as $(U_n^{*,\sigma}(x), V_n^{*,\sigma}(x))$. $D^n(\varepsilon, \sigma)$ is called the normal n-layered solution or simply n-layer solution.*

3.2. Asymptotic Behaviors of Critical Eigenvalues and Eigenfunctions of $L^{\varepsilon,\sigma}$

We shall study the stability properties of the normal n-layered solutions $D^n(\varepsilon, \sigma)$ in the following three subsections. Since our model is a semilinear parabolic system, the stability is determined by the spectrum of the following linearized operator at $D^n(\varepsilon, \sigma)$

$$\mathcal{L}^{\varepsilon,\sigma} \begin{pmatrix} w \\ z \end{pmatrix} \equiv \begin{pmatrix} L^{\varepsilon,\sigma} & f_v^{\varepsilon,\sigma} \\ g_u^{\varepsilon,\sigma} & M^{\varepsilon,\sigma} \end{pmatrix} \begin{pmatrix} w \\ z \end{pmatrix} = \lambda \begin{pmatrix} w \\ z \end{pmatrix}, \tag{3.12a}$$

$$(w, z)^t \in \{H^2(I) \cap H_N^1(I)\}^2$$

where

$$L^{\varepsilon,\sigma} \equiv \varepsilon^2 \frac{d^2}{dx^2} + f_u^{\varepsilon,\sigma} \tag{3.12b}$$

$$M^{\varepsilon,\sigma} \equiv \frac{1}{\sigma}\frac{d^2}{dx^2} + g_v^{\varepsilon,\sigma}, \tag{3.12c}$$

and all partial derivatives are evaluated at $D^n(\varepsilon, \sigma) = (u^n(x; \varepsilon, \sigma), v^n(x; \varepsilon, \sigma))$, i.e., $f_u^{\varepsilon,\sigma} \equiv f_u(u^n(x; \varepsilon, \sigma), v^n(x; \varepsilon, \sigma))$ and so on.

A naive way to find the spectral behavior of (3.12) as $\varepsilon \downarrow 0$ is to put $\varepsilon = 0$ in (3.12), however it turns out that the resulting system is exactly the same as (2.22) in Section 2. Namely just the formal limit of (3.12) tells us only the behavior of noncritical eigenvalues which govern the behavior of solutions in outer region, and does not inherit any information from layer part. As we saw in Section 2, the most subtle part of spectral analysis comes from the layer part. The reason for this is that it is related to neutral (zero) eigenvalue of *translation invariance*. More precisely, rewriting (3.12) by using a stretched coordinate $y = (x - x_i^*)/\varepsilon$ at any layer position x_i^* and taking a formal limit of $\varepsilon \downarrow 0$, we see from Theorem 3.3 that (3.12) restricted to the i-th subinterval $I_i \equiv ((i-1)/n, i/n)$ becomes

$$\frac{d^2}{dy^2}\tilde{w} + f_u(\bar{u}^*, v^*)\tilde{w} + f_v(\bar{u}^*, v^*)\tilde{z} = \lambda\tilde{w}$$

$$y \in \mathbf{R} \tag{3.13}$$

$$\frac{d^2}{dy^2}\tilde{z} = 0.$$

It follows from Remark 3.7 that (3.13) has zero eigenvalue with $(\tilde{w}, \tilde{z}) = (\frac{d\bar{u}^*}{dy}, 0)$. This observation suggests that (3.12) has an eigenvalue which approaches zero as $\varepsilon \downarrow 0$ associated with each layer. In fact the number of these *critical* eigenvalues (i.e., those which tend to zero as $\varepsilon \downarrow 0$) of (3.12) will be proved to be exactly equal to the number of layers ($= n$) and their precise behavior will be clarified by solving the SLEP matrix which eventually leads us to the main result (Theorem 3.25).

In view of (3.13), we see that each zero eigenvalue coincides with that of the limiting (stretched) Sturm-Liouville operator, hence $L^{\varepsilon,\sigma}$ itself is expected to have n eigenvalues going to zero when $\varepsilon \downarrow 0$. In fact, we shall show in Lemma 3.10 that there are n *positive* critical eigenvalues of $L^{\varepsilon,\sigma}$, and the rest of the spectrum is strictly bounded away from zero. The asymptotic behavior of these critical eigenvalues will be investigated in Lemma 3.15. Note that we use the word "critical" both for eigenvalues of $L^{\varepsilon,\sigma}$ and the full system $\mathcal{L}^{\varepsilon,\sigma}$.

The supports of eigenfunctions associated with the critical eigenvalues of $L^{\varepsilon,\sigma}$ are concentrated on layer positions when $\varepsilon \downarrow 0$, in fact, after an appropriate scaling, each eigenfunction tends to a combination of Dirac's point mass distribution on layer positions (Lemma 3.22). This seems very singular, however, we will see that it is equivalent to say that the stretched eigenfunction at each layer position approaches a constant multiple of the principal eigenfunction of the limiting Sturm-Liouville problem (3.11).

The next lemma tells us the number of critical eigenvalues of the Sturm-Liouville problem:

$$L^{\varepsilon,\sigma}\phi = \zeta\phi \qquad \text{on} \quad I$$

$$\tag{3.14}$$

$$\phi \in H^2(I) \cap H_N(I), \qquad \| \phi \|_{L^2(I)} = 1.$$

We denote the complete orthonormal set of (3.14) by $\{\zeta_i^{\varepsilon,\sigma}, \phi_i^{\varepsilon,\sigma}\}$. Throughout this subsection we omit the superscript σ like $L^\varepsilon, \zeta_i^\varepsilon, \phi_i^\varepsilon$ in the proofs of lemmas.

Lemma 3.10. *The first n eigenvalues of $L^{\varepsilon,\sigma}$ tend to zero as $\varepsilon \downarrow 0$, and the rest of the eigenvalues of $L^{\varepsilon,\sigma}$ is strictly negative up to $\varepsilon = 0$.*

Proof. We first consider the behavior of the principal eigenvalue of (3.14). Since a normal n-layered solution is obtained by folding up a mono-layered solution n-times, the principal eigenfunction of L^ε is also obtained by folding that of L^ε restricted to the first subinterval $I_1 \equiv (0, 1/n)$

$$L^\varepsilon \phi_{01}^\varepsilon = \zeta_0^\varepsilon \phi_{01}^\varepsilon \qquad \text{on } I_1$$

$$\phi_{01}^\varepsilon \in H^2(I_1) \cap H_N(I_1), \qquad \| \phi_{01}^\varepsilon \|_{L^2(I)} = \frac{1}{n},$$

(3.15)

where ζ_0^ε is the principal eigenvalue of (3.14) and ϕ_{01}^ε denotes the restriction of the associated principal eigenfunction to I_1. Using a stretched coordinate $y = (x - x_1^*)/\varepsilon$, (3.15) becomes the following with a new normalization :

$$\frac{d^2}{dy^2}\tilde{\phi}_{01}^\varepsilon + \tilde{f}_u^\varepsilon \tilde{\phi}_{01}^\varepsilon = \zeta_0^\varepsilon \tilde{\phi}_{01}^\varepsilon, \qquad y \in \tilde{I}_1,$$

$$\tilde{\phi}_{01}^\varepsilon \in H^2(\tilde{I}_1) \cap H_N(\tilde{I}_1), \qquad \| \tilde{\phi}_{01}^\varepsilon \|_{L^2(\tilde{I}_1)} = 1$$

(3.16)

where $\tilde{I}_1 \equiv (-\ell/\varepsilon, r/\varepsilon)$ with $\ell = x_1^*$, $r = 1/n - x_1^*$, and $\tilde{\phi}_{01}^\varepsilon (> 0)$ is the normalized stretched principal eigenfunction. In view of Corollary 3.6, we see that $\lim_{\varepsilon \downarrow 0} \tilde{f}_u^\varepsilon = f_u(\tilde{u}^*, v^*)$ in $C_{c.u.}^2(\mathbf{R})$-sense. Recalling the behavior in outer region (see Theorem 3.3), we can find a finite interval I_0 and positive constants μ and γ such that

$$-\gamma < \tilde{f}_u^\varepsilon < -\mu < 0 \qquad y \in \tilde{I}_1 \backslash I_0$$

(3.17)

holds, where I_0, μ, γ are independent of $(\varepsilon, \sigma) \in Q^n$.

First we give a lower bound for ζ_0^ε. It is clear that ζ_0^ε is characterized by

$$\zeta_0^\varepsilon \equiv \sup_{\tilde{\phi} \in H_N^1(\tilde{I}_1), \ \| \tilde{\phi} \|_{L^2(\tilde{I}_1)} = 1} \left(-\langle \tilde{\phi}_y, \tilde{\phi}_y \rangle + \langle \tilde{f}_u^\varepsilon \tilde{\phi}_y, \tilde{\phi}_y \rangle \right).$$

(3.18a)

We know from Remark 3.7 that

$$0 = \sup_{\tilde{\phi} \in H^1(\mathbf{R}), \ \| \tilde{\phi} \|_{L^2(\mathbf{R})} = 1} \left(-\langle \tilde{\phi}_y, \tilde{\phi}_y \rangle + \langle \tilde{f}_u^* \tilde{\phi}_y, \tilde{\phi}_y \rangle \right)$$

(3.18b)

which is attained by $\hat{\phi}_0 \left(= \frac{\gamma^*}{c^*} \frac{d}{dy}\tilde{u}^* \right)$. Taking a constant multiple of $\hat{\phi}_0^*$ (restricted to $\tilde{I}_1$) as a test function for (3.18a), and using the facts that $\lim_{\varepsilon \downarrow 0} \tilde{f}_u^\varepsilon = \tilde{f}_u^*$ in $C_{c.u.}^2(\mathbf{R})$-sense and $\frac{d}{dy}\tilde{u}^*$ decays exponentially as $|y| \to \infty$, we see that, for arbitrary small $\delta > 0$, there exists $\varepsilon_\delta > 0$ such that

$$\zeta_0^\varepsilon > -\delta \quad \text{for} \quad 0 < \varepsilon < \varepsilon_\delta$$

holds. Also it is clear from (3.18a) that $\zeta_0^\varepsilon < \max_y \tilde{f}_u^\varepsilon$. These inequalities imply that

$$q^\varepsilon(y) \equiv \tilde{f}_u^\varepsilon - \zeta_0^\varepsilon \tag{3.19}$$

also satisfies the inequality (3.17) for small ε. Using this property, we can show the exponentially decaying property of $\tilde{\phi}_{01}^\varepsilon$:

$$|\tilde{\phi}_{01}^\varepsilon| \le C \exp(-C_1|y|), \tag{3.20}$$

which will be proved in Lemma 3.11 in more general setting. Multiplying $\tilde{\phi}_{01}^\varepsilon$ on both sides of (3.16) and integrating over $\bar{I}_1$, we see from (3.19) that $\| \tilde{\phi}_{01}^\varepsilon \|_{H^1(\bar{I}_1)} <$ M independently of ε, which implies via Sobolev imbedding theorem that $\tilde{\phi}_{01}^\varepsilon$ has a convergent subsequence $\tilde{\phi}_{01}^\varepsilon$ (hereafter we use the same notation for subsequences) on any compact subset $K(\subset \mathbf{R})$ in $C^0(K)$-topology. Using (3.16) again, this becomes a convergent sequence in $C^2(K)$-topology. Using a diagonal argument on an expanding sequence of compact intervals to $\mathbf{R}$ and recalling the boundedness of ζ_0^ε, we can find a convergent subsequence $(\tilde{\phi}_{01}^\varepsilon, \zeta_0^\varepsilon)$ in $C_{c.u.}^2(\mathbf{R}) \times \mathbf{R}$-topology. The limiting function denoted by $(\tilde{\Phi}_{01}^*, \zeta_0^*)$ satisfies

$$\frac{d^2}{dy^2}\tilde{\Phi}_{01}^* + \tilde{f}_u^* \tilde{\Phi}_{01}^* = \zeta_0^* \tilde{\Phi}_{01}^*. \tag{3.21}$$

We shall show that

$$\lim_{\varepsilon \downarrow 0}(\tilde{\phi}_{01}^\varepsilon, \zeta_0^\varepsilon) = (\tilde{\phi}_{01}^*, \zeta_0^*) = (\hat{\phi}_{01}^*, 0) \quad \text{in} \quad C^2(\mathbf{R}) \times \mathbf{R}\text{-sense}. \tag{3.22}$$

Here we use the convention of Remark 3.5 for $\tilde{\phi}_{01}^\varepsilon$. First, because of (3.20), we see that $\tilde{\phi}_{01}^\varepsilon \not\equiv 0$ and satisfies (3.21), moreover $\tilde{\phi}_{01}^* > 0$ since all $\tilde{\phi}_{01}^\varepsilon$ are positive principal eigenfunctions. On the other hand, we know (see Remark 3.7) that (3.21) has zero as the principal eigenvalue with $\hat{\phi}_0^*$ being the associated eigenfunction. Hence, because of the simplicity of the principal eigenvalue, we obtain (3.22) in $C_{c.u.}^2(\mathbf{R}) \times \mathbf{R}$-sense for a chosen subsequence. Apparently the limiting function does not depend on the choice of the subsequence, and taking into account (3.20), we can conclude that $(\tilde{\phi}_{01}^\varepsilon, \zeta_0^\varepsilon)$ itself converges to $(\hat{\phi}_0^*, 0)$ in $C^2(\mathbf{R}) \times \mathbf{R}$-topology, which completes the proof of (3.22).

All the above discussions remain valid when we change the boundary conditions to Dirichlet ones (with replacing $H_N^1(\bar{I}_1)$ by $H_0^1(\bar{I}_1)$ in (3.16)). In particular, the principal eigenvalue ζ_{01D}^ε of $(3.15)_D$ (or $(3.16)_D$) also satisfies

$$\lim_{\varepsilon \downarrow 0}\zeta_{01D}^\varepsilon = 0, \tag{3.23}$$

where the subscript D represents that the concerned quantity is considered under Dirichlet boundary conditions. Note that, by comparison theorem,

$$\zeta_{01D}^\varepsilon < \zeta_0^\varepsilon \qquad \text{for} \quad \varepsilon > 0. \tag{3.24}$$

We denote by ϕ_{01D}^ε the principal eigenfunction of $(3.15)_D$.

Now we return to the problem (3.14) for n-layered solution. Flipping ϕ_{01}^ε of (3.15) n times in an even way, the resulting function (denoted by ϕ_0^ε) defined on I, becomes a principal eigenfunction (i.e., nodal zero) of (3.14). Note that the eigenvalue ζ_0^ε remains the same as before by folding operation. On the other hand, by flipping ϕ_{01D}^ε n-times in an odd way, we obtain the eigenfunction (denoted by $\phi_{n-1,D}^\varepsilon$) of $(3.14)_D$ which has $n-1$ nodal zeros inside of I. Hence ζ_{01D}^ε becomes the n-th eigenvalue $\zeta_{n-1,D}^\varepsilon$ of L^ε under Dirichlet boundary conditions. Making use of the comparison theorem and the nodal property of Sturm-Liouville operator, we see that

$$\zeta_{n-1,D}^\varepsilon = \zeta_{01D}^\varepsilon < \zeta_{n-1}^\varepsilon < \cdots < \zeta_0^\varepsilon$$

holds, namely the eigenvalues $\zeta_1^\varepsilon, \cdots, \zeta_{n-1}^\varepsilon$ are sandwiched by ζ_0^ε and ζ_{01D}^ε. The asymptotic behaviors (3.22) and (3.23) lead to the first part of Lemma 3.10. $\square$

As for the second part, first note that the second eigenvalue ζ_{11}^ε (nodal one) of (3.15) is equal to the $(n+1)$-th eigenvalue ζ_n^ε (nodal n) of (3.14). Since the principal eigenvalue ζ_0^ε of (3.15), which is simple up to $\varepsilon = 0$, converges to zero as $\varepsilon \downarrow 0$, ζ_{11}^ε and hence ζ_n^ε must be strictly negative for small ε. This completes the proof of Lemma 3.10.

In order to know the precise asymptotic behaviors of $\zeta_i^{\varepsilon,\sigma}$ ($i = 0, \cdots, n-1$), we need to find the asymptotic forms of $\phi_i^{\varepsilon,\sigma}$ ($i = 0, \cdots, n-1$). As we saw in the proof of Lemma 3.10, the stretched function $\tilde\phi_{01}^{\varepsilon,\sigma}$ (on unit subinterval) converges to the principal eigenfunction of (3.11). Noting the relation $\int_{I_1} \left|\dfrac{\phi}{\sqrt\varepsilon}\right|^2 dx = \int_{\tilde I_1} |\tilde\phi|^2 dy$ between a function $\phi(x)(x \in I_1)$ and its stretched one $\tilde\phi(y)(y \in \tilde I_1)$, we see that the L^2-normalized function $\phi_{01}^{\varepsilon,\sigma}(x)$ has a sharp peak of height $O(1/\sqrt\varepsilon)$, and decays quickly outside of this peak as $\varepsilon \downarrow 0$. Hence it is indispensable to use *stretching* to characterize the asymptotic behavior of critical eigenfunctions. Since there are n layers, it is convenient to introduce the following notation.

$$\phi_{ij}^{\varepsilon,\sigma} \equiv \phi_i^{\varepsilon,\sigma}\big|_{I_j} \text{ ; restriction to the subinterval } I_j \text{ of the } i\text{-th eigenfunction,}$$

$$\tilde\phi_{ij}^{\varepsilon,\sigma} = \tilde\phi_{ij}^{\varepsilon,\sigma}(y) \equiv \phi_{ij}^{\varepsilon,\sigma}(x_j^* + \varepsilon y); \text{ stretched function of } y \in \tilde I_j,$$

where $y = (x - x_j^*)/\varepsilon$ and $\tilde I_j$ denotes $(-\ell/\varepsilon, r/\varepsilon)$ or $(-r/\varepsilon, \ell/\varepsilon)$. It holds that

$$\int_I |\phi_i^{\varepsilon,\sigma}|^2 dx = \sum_{j=1}^n \int_{I_j} |\phi_{ij}^{\varepsilon,\sigma}|^2 dx = \sum_{j=1}^n \int_{\tilde I_j} |\sqrt\varepsilon \tilde\phi_{ij}^{\varepsilon,\sigma}|^2 dy. \tag{3.25}$$

We define $\hat\phi_{ij}^{\varepsilon,\sigma}$ by

$$\hat\phi_{ij}^{\varepsilon,\sigma}(y) \equiv \sqrt\varepsilon \tilde\phi_{ij}^{\varepsilon,\sigma}, \qquad y \in \tilde I_j. \tag{3.26}$$

We call $\hat\phi_{ij}^{\varepsilon,\sigma}$ the *normalized j-th stretched function* of $\phi_i^{\varepsilon,\sigma}$. It is obvious from (3.25) that

$$\int_I |\phi_i^{\varepsilon,\sigma}|^2 dx = \sum_{j=1}^n \int_{\tilde I_j} |\hat\phi_{ij}^{\varepsilon,\sigma}|^2 dy \tag{3.27}$$

A key result for the critical eigenfunctions is that, for any i and j, $\hat{\phi}_{ij}^{\varepsilon,\sigma}$ converges to a constant multiple of $\frac{d}{dy}\bar{u}^*$ in $C^2(\mathbf{R})$-sense as $\varepsilon \downarrow 0$ (Lemma 3.12). Namely, if we look at a critical eigenfunction in a stretched form, it is very close to the derivative of the stretched layer function $\bar{u}^*(y)$ of Theorem 3.3. To this end, we need the following lemma.

Lemma 3.11 (Exponential decaying property). *Let $\hat{\phi}_{ij}^{\varepsilon,\sigma}$ be the normalized j-th stretched eigenfunction of $\phi_i^{\varepsilon,\sigma}$ ($0 \le i \le n-1$, $1 \le j \le n$). Then there is a finite interval I_0, which is independent of ε, σ, i, and j, such that*

$$\left|\frac{d^k}{dy^k}\hat{\phi}_{ij}^{\varepsilon,\sigma}\right| \le C\exp(-C_1|y|), \qquad y \in \tilde{I}_j\backslash I_0 \tag{3.28}$$

hold for $k = 0, 1, 2$, where C and C_1 are positive constants independent of ε, σ, i, and j.

Proof. In view of Lemma 3.10 and (3.19), we have for small ε

$$-\gamma < \tilde{f}_u^\varepsilon - \zeta_i^\varepsilon < -\mu < 0, \qquad y \in \tilde{I}_j\backslash I_0, \tag{3.29}$$

where μ, γ, and I_0 are same as in (3.17). We decompose $\tilde{I}_j$ into three parts (see Figure 3.1): $\tilde{I}_j = \tilde{I}_{j,\ell} \cup I_0 \cup \tilde{I}_{j,r}$.

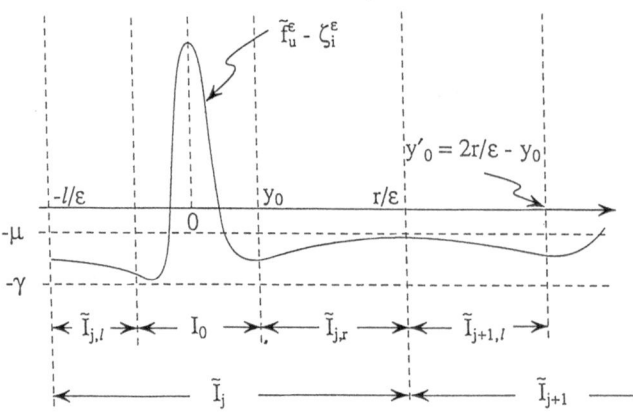

Figure 3.1

Without loss of generality, we prove (3.28) on the interval $\tilde{I}_{j,r} \equiv (y_0, r/\varepsilon)$. We consider the problem on the extended interval $\tilde{I}_R \equiv \tilde{I}_{j,r} \cup \tilde{I}_{j+1,\ell}$. If $\tilde{I}_{j,r}$ is the right-end interval, we extend everything to the right in an even manner. Note that $\tilde{f}_u^\varepsilon - \zeta_i^\varepsilon$ has reflectional symmetry at $y = r/\varepsilon$ and (3.29) holds on $\tilde{I}_R$. By using (3.29), it is not difficult to show that there exist two linearly independent solutions $\psi_+(y)$ and $\psi_-(y)$ of

$$\frac{d^2}{dy^2}\tilde{\phi} + (\tilde{f}_u^\varepsilon - \zeta_i^\varepsilon)\tilde{\phi} = 0 \tag{3.30}$$

on $\tilde{I}_R$ satisfying

$$\psi_+(y_0) = 1, \qquad \frac{d}{dy}\psi_+(y_0') = 0, \tag{3.31a}$$

$$C_2 \exp(-\gamma y) \le \frac{d^k}{dy^k}\psi_+(y) \le C_3 \exp(-\mu y), \qquad k = 0, 1, 2, \tag{3.31b}$$

$$\psi_-(y) \equiv \psi_+(2r/\varepsilon - y), \qquad y \in \tilde{I}_R, \tag{3.31c}$$

where $y_0' = 2r/\varepsilon - y_0$, C_2 and C_3 are positive constants independent of ε, σ, i and j. Here we use the fact that $f_u^\varepsilon - \zeta_i^\varepsilon$ has reflectional symmetry at $y = r/\varepsilon$ which implies that $\psi_-(y)$ defined by (3.31c) becomes a solution of (3.30). It is convenient to introduce the following pair of linearly independent solutions:

$$\begin{aligned} \Psi_s &\equiv \frac{1}{2}(\psi_+ + \psi_-) \\ \Psi_a &\equiv \frac{1}{2}(\psi_+ - \psi_-) \end{aligned} \tag{3.32}$$

where Ψ_s (resp. Ψ_a) is an even (resp. odd) function at $y = r/\varepsilon$. $\hat{\phi}_{ij}^\varepsilon$ can be expressed by

$$\hat{\phi}_{ij}^\varepsilon = c_s \Psi_s + c_a \Psi_a \qquad \text{on } \tilde{I}_R. \tag{3.33}$$

Since $\| \phi_i^\varepsilon \|_{L^2(I)} = 1$, we see from (3.27), (3.33) and $\langle \Psi_s, \Psi_a \rangle_{\tilde{I}_R} = 0$ that

$$\int_{\tilde{I}_R} |\hat{\phi}_{ij}^\varepsilon|^2 dy = c_s^2 \| \Psi_s \|_{L^2(\tilde{I}_R)}^2 + c_a^2 \| \Psi_a \|_{L^2(\tilde{I}_R)}^2 \le 1.$$

It is clear from (3.31b), that

$$m_1 < \| \Psi_s \|_{L^2(\tilde{I}_R)}, \qquad \text{and} \qquad \| \Psi_a \|_{L^2(\tilde{I}_R)} < m_2,$$

where $m_i(i = 1, 2)$ are positive constants independent of ε, σ, i and j. These two inequalities lead to the following

$$|c_s|, \quad |c_a| \le M,$$

where M does not depend on parameters. In terms of $\psi_\pm$, $\hat{\phi}_{ij}^\varepsilon$ can be written as $\hat{\phi}_{ij}^\varepsilon = (c_s + c_a)\psi_+/2 + (c_s - c_a)\psi_-/2$, where the coefficients are bounded due to the above estimate. This implies the required estimate for $\hat{\phi}_{ij}^\varepsilon$, since we see from (3.31) that ψ_+ (and its derivatives) decays exponentially and the contribution of ψ_- on $\tilde{I}_{j,r}$ is exponentially small. $\square$

Now we are ready to prove the following.

Lemma 3.12 (Precompactness of $\hat{\phi}_{ij}^\varepsilon$). *There exists a subsequence* $\{\hat{\phi}_{ij}^{\varepsilon_m,\sigma}\}_{m=1}^\infty$ *with* $\lim_{m\uparrow\infty} \varepsilon_m = 0$ *from* $\{\hat{\phi}_{ij}^{\varepsilon,\sigma}\}$ *such that*

$$\lim_{m\uparrow\infty} \hat{\phi}_{ij}^{\varepsilon_m,\sigma} = \kappa_j^i \hat{\phi}_L^* \qquad \text{in } C^2(\mathbf{R})\text{-sense (recall Remark 3.5)} \tag{3.34}$$

hold simultaneously for all i and $j(0 \leq i \leq n - 1, 1 \leq j \leq n)$. The resulting vectors $\{c^* \mathbf{k}^i\}_{i=0}^{n-1}$ defined by

$$c^* \mathbf{k}^i \equiv c^* (\kappa_1^i, \cdots, \kappa_n^i)$$

$$c^* \equiv \|\hat{\phi}_L^*\|_{L^2(\mathbf{R})}$$

form an othonormal set in $\mathbf{R}^n$, i.e.,

$$(c^*)^2 \sum_{j=1}^{n} \kappa_j^\ell \kappa_j^k = \delta_{\ell k}, \qquad \ell, k \in \{0, 1, \cdots, n-1\}, \tag{3.35}$$

where $\delta_{\ell k}$ denotes the Kronecker's δ.

Remark 3.13. The coefficients κ_j^i in Lemma 3.12 except $i = 0$ may depend on the choice of subsequence, however it does not affect later discussions. In fact we will see in Section 3.3 that the final form (3.75) of the SLEP system is independent of the choice of subsequence. It is conjectured that $\hat{\phi}_{ij}^{\varepsilon,\sigma}$ has a unique limit and κ_j^i does not depend on subsequences.

Proof. Using Lemma 3.11, the proof of (3.34) for a fixed i and j proceeds in a quite similar way as in the proof of Lemma 3.10 where $(\tilde{\phi}_{01}^\varepsilon, \zeta_0^\varepsilon)$ was shown to converge to $(\hat{\phi}_0^*, 0)$ in $C^2(\mathbf{R}) \times \mathbf{R}$-sense when $\varepsilon \downarrow 0$. However, because of the lack of knowledge about the convergence of $\| \hat{\phi}_{ij}^\varepsilon \|_{L^2(I_j)}$ ($i \geq 1$) as $\varepsilon \downarrow 0$ (although it is uniformly bounded), we have to choose a subsequence, and this cause the above anbiguity, i.e., κ_j^i may depend on the choice of subsequence. Using a diagonal argument for i and j, we can find a subsequence ε_m such that (3.34) holds for all i and j. The orthogonality (3.35) comes from that of the eigenfunctions of the Sturm-Liouville operator, i.e., $\langle \phi_i^\varepsilon, \phi_j^\varepsilon \rangle = \delta_{ij}$. $\square$

Corollary 3.14.

$$\int_I |\phi_i^{\varepsilon,\sigma}| dx = L_i(\varepsilon, \sigma)\sqrt{\varepsilon},$$

where $L_i(\varepsilon, \sigma)$ is a positive continuous function in Q^n and satisfies

$$L_i^* \equiv \lim_{m \uparrow \infty} L_i(\varepsilon_m, \sigma) = \sum_{j=1}^{n} |\kappa_j^i|. \tag{3.36}$$

Proof. On each subinterval I_j, we have

$$\int_{I_j} \phi_{ij}^\varepsilon dx = \sqrt{\varepsilon} \int_{I_j} \sqrt{\varepsilon} \tilde{\phi}_{ij}^\varepsilon dy$$

$$= \sqrt{\varepsilon} \int_{I_j} \hat{\phi}_{ij}^\varepsilon dy.$$

On the other hand, it follows from (3.34) and $\int_{\mathbf{R}} \hat{\phi}_L^* = 1$ that

$$
\lim_{\varepsilon \downarrow 0} \int_{I_j} |\hat{\phi}_{ij}^\varepsilon| dy = |\kappa_j^i| \int_{\mathbf{R}} \hat{\phi}_L^* dy \tag{3.37}
$$

$$
= |\kappa_j^i|.
$$

Hence

$$
\int_{I_j} |\phi_i^\varepsilon| dx = \sqrt{\varepsilon} \sum_{j=1}^{n} \int_{\bar{I}_j} |\hat{\phi}_{ij}^\varepsilon| dy.
$$

We define $L_i(\varepsilon, \sigma)$ by

$$
L_i(\varepsilon, \sigma) = \sqrt{\varepsilon} \sum_{j=1}^{n} \int_{\bar{I}_j} |\hat{\phi}_{ij}^\varepsilon| dy.
$$

Since the integral term behaves continuously for $\varepsilon > 0$, so does $L_i(\varepsilon, \sigma)$ in Q^n. the equality (3.36) is a direct consequence of (3.37). □

Lemma 3.15 (Asymptotic Behaviors of Eigenvalues of $L^{\varepsilon,\sigma}$). *Let $\{\zeta_i^{\varepsilon,\sigma}, \phi_i^{\varepsilon,\sigma}\}_{i=0}^{\infty}$ be the complete orthonormal set of eigenvalues and eigenfunctions of*

$$
L^{\varepsilon,\sigma} \equiv \left(\varepsilon^2 \frac{d^2}{dx^2} + f_u^{\varepsilon,\sigma} \right) \phi = \zeta \phi,
$$

subject to Neumann boundary conditions, where $f_u^{\varepsilon,\sigma}$ denotes the partial derivative f_u evaluated at the normal n-layered solution. The first n eigenvalues $\zeta_0^{\varepsilon,\sigma}, \cdots, \zeta_{n-1}^{\varepsilon,\sigma}$ ($\zeta_0^{\varepsilon,\sigma} > \cdots > \zeta_{n-1}^{\varepsilon,\sigma}$) are critical eigenvalues of $L^{\varepsilon,\sigma}$, which are positive for $\varepsilon > 0$ and satisfy the asymptotic formula as $\varepsilon \downarrow 0$ (see Figure 3.2)

$$
\zeta_i^{\varepsilon,\sigma} = \hat{\zeta}_i(\varepsilon, \sigma)\varepsilon\sigma + e_i(\varepsilon, \sigma). \tag{3.38}
$$

Here $\hat{\zeta}_i$ and $e_i (i = 0, \cdots, n-1)$ are positive continuous functions in Q^n (see Corollary 3.9 and Theorem 3.3) which are continuously extendable to $\varepsilon = 0$ and satisfy the following:

$$
\hat{\zeta}^{*,\sigma} \equiv \lim_{\varepsilon \downarrow 0} \hat{\zeta}_i(\varepsilon, \sigma) \tag{3.39}
$$

$$
= \frac{1}{n} \left(\frac{\gamma^*}{c^*} \right)^2 \frac{dJ}{dv}(v^*) \int_0^{x_1^*(\frac{\sigma}{n^2})} g \left(U^{*,\sigma/n^2}, V^{*,\sigma/n^2} \right) dx > 0,
$$

$$
|e_i(\varepsilon, \sigma)| \leq C \exp \left(-\frac{\gamma}{\varepsilon} \right), \tag{3.40}
$$

where γ^* and c^* are positive constants defined in Remark 3.7, $\left(U^{*,\sigma/n^2}, V^{*,\sigma/n^2} \right)$ is the basic pattern in Theorem 3.3 with replacing σ by σ/n^2, and C and γ are positive constants independent of $(\varepsilon, \sigma) \in Q^n$ and i. Note that the asymptotic limit $\hat{\zeta}^{*,\sigma}$ does not depend on i $(0 \leq i \leq n - 1)$. The rest of the eigenvalues $\zeta_i^{\varepsilon,\sigma} (i \geq n)$ are negative and uniformly bounded away from zero with respect to small ε, namely, it holds that

$$0 > -\Delta^* > \zeta_n^{\varepsilon,\sigma} > \zeta_{n+1}^{\varepsilon,\sigma} > \cdots$$

for small ε, where Δ^* is a positive constant independent of $(\varepsilon, \sigma) \in Q^n$.

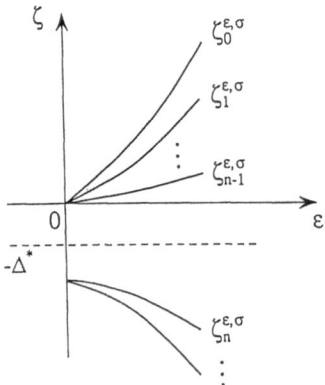

Figure 3.2.

Remark 3.16. *Lemma 3.15 also holds under homogeneous Dirichlet boundary conditions without essential changes. In particular, the asymptotic limit $\zeta^{*,\sigma}$ remains the same as the Neumann case.*

Proof. Let $\phi_i^\varepsilon (0 \le i \le n-1)$ be the L^2-normalized i-th eigenfunction of L^ε, and recall that

$$\hat{\phi}_{ij}^\varepsilon(y) \equiv \sqrt{\varepsilon}\phi_i^\varepsilon(x_j^* + \varepsilon y), \qquad y \in \tilde{I}_j \equiv \left(-\frac{\ell}{\varepsilon}, \frac{r}{\varepsilon}\right),$$

i.e., the stretched function of ϕ_i^ε with center at x_j^* on j-th subinterval $(1 \le j \le n)$, and

$$\int_{I_j} |\phi_i^\varepsilon|^2 dx = \int_{\tilde{I}_j} |\hat{\phi}_{ij}^\varepsilon|^2 dy.$$

It is clear that $\hat{\phi}_{ij}^\varepsilon$ satisfies

$$\frac{d^2}{dy^2}\hat{\phi}_{ij}^\varepsilon + \tilde{f}_u^\varepsilon \hat{\phi}_{ij}^\varepsilon = \zeta_i^\varepsilon \hat{\phi}_{ij}^\varepsilon \qquad \text{on } \tilde{I}_j \tag{3.41}$$

and from (3.34)

$$\lim_{\varepsilon \downarrow 0} \hat{\phi}_{ij}^\varepsilon = \kappa_j^i \hat{\phi}_{ij}^\varepsilon \qquad \text{in } C^2(\mathbf{R})\text{-sense} \tag{3.42}$$

holds for an appropriately chosen subsequence. Here we keep the notation ε instead of ε_m, since the final result (3.39) does not depend on the choice of subsequence. We take j such that $\kappa_j^i \ne 0$. On the other hand, the y-derivative of the stretched normal n-layered solution satisfies

$$\frac{d^2}{dy^2}\tilde{u}_y^\varepsilon + \tilde{f}_u^\varepsilon \tilde{u}_y^\varepsilon = -\tilde{f}_v^\varepsilon \tilde{v}_y^\varepsilon \qquad y \in \tilde{I}_j. \tag{3.43}$$

Multiplying $\tilde{u}_y^\varepsilon$ on both sides of (3.41) and integrating by parts twice, we have

$$\langle \hat{\phi}_{ij}^\varepsilon, -\tilde{f}_v^\varepsilon \tilde{v}_y^\varepsilon \rangle - \hat{\phi}_{ij}^\varepsilon \frac{d}{dy} \tilde{u}_y^\varepsilon \bigg|_{-\ell/\varepsilon}^{r/\varepsilon} = \zeta_i^\varepsilon \langle \hat{\phi}_{ij}^\varepsilon, \tilde{u}_y^\varepsilon \rangle. \tag{3.44}$$

Here we use (3.43) and the fact that $\tilde{u}_y^\varepsilon = 0$ at $y = -\ell/\varepsilon, r/\varepsilon$. Using Lemma 3.11, we see that

$$|\text{the second term on the left-hand side of (3.44)}| \le C \exp\left(-\frac{\gamma}{\varepsilon}\right), \tag{3.45}$$

where C, γ are positive constants independent of ε and σ. In order to compute the first term of (3.44), first note that $\tilde{v}^\varepsilon$ satisfies

$$\frac{1}{\sigma}\frac{d^2}{dy^2}\tilde{v}^\varepsilon + \varepsilon^2 g(\tilde{u}^\varepsilon, \tilde{v}^\varepsilon) = 0. \tag{3.46}$$

Integrating (3.46) from $-\ell/\varepsilon$ to y and using $\tilde{v}_y^\varepsilon(-\ell/\varepsilon) = 0$, we have

$$\tilde{v}_y^\varepsilon = -\sigma\varepsilon\tilde{\theta}_j^\varepsilon(y), \tag{3.47}$$

where

$$\tilde{\theta}_j^\varepsilon(y) \equiv \varepsilon \int_{-\ell/\varepsilon}^y g(\tilde{u}^\varepsilon, \tilde{v}^\varepsilon)dy.$$

Here $\tilde{\theta}_j^\varepsilon(y)$ can be obtained by stretching the following function at $x = x_j^*$:

$$\theta_j^\varepsilon(x) \equiv \int_{-\ell+x_j^*}^x g(u^\varepsilon, v^\varepsilon)dx.$$

θ_j^ε is C^1 for $\varepsilon > 0$ and its C^1-norm is uniformly bounded with respect to ε, and hence we see from Remark 3.4 that $\tilde{\theta}_j^\varepsilon(y)$ converges to a constant function as $\varepsilon \downarrow 0$. More precisely,

$$\lim_{\varepsilon\downarrow 0} \tilde{\theta}_j^\varepsilon(y) = \int_{-\ell+x_j^*}^{x_j^*} g(U_n^*, V_n^*)dx. \tag{3.48}$$

holds in $C_{c.u.}^2(\mathbf{R})$-sense, where (U_n^*, V_n^*) is the reduced solution of the normal n-layered solution. In view of Lemma 3.12 and Remark 3.7, we see that

$$\lim_{\varepsilon\downarrow 0} \hat{\phi}_{ij}^\varepsilon = \kappa_j^i \hat{\phi}_L^* = \gamma^* \frac{d}{dy}\tilde{u}^*$$

$$\tag{3.49}$$

$$\lim_{\varepsilon\downarrow 0} \tilde{u}_y^\varepsilon = \frac{c^*}{\gamma^*}\hat{\phi}_0^*$$

hold in $C^2_{c.u.}(\mathbf{R})$-sense. Substituting (3.47) into (3.44), ζ^ε_i is expressed by

$$\zeta^\varepsilon_i = \hat{\xi}_i(\varepsilon, \sigma)\sigma\varepsilon + e_i(\varepsilon, \sigma), \tag{3.50a}$$

where

$$\hat{\xi}_i(\varepsilon, \sigma) \equiv \frac{\langle \tilde{f}^\varepsilon_v \hat{\phi}^\varepsilon_{ij}, \tilde{\theta}^\varepsilon_j \rangle}{\langle \hat{\phi}^\varepsilon_{ij}, \tilde{u}^\varepsilon_y \rangle} \tag{3.50b}$$

$$e_i(\varepsilon, \sigma) \equiv -\hat{\phi}^\varepsilon_{ij}\tilde{u}^\varepsilon_{yy}\Big|^{r/\varepsilon}_{-\ell/\varepsilon}/\langle \hat{\phi}^\varepsilon_{ij}, \tilde{u}^\varepsilon_y \rangle. \tag{3.50c}$$

It is clear from (3.45) that $e_i(\varepsilon, \sigma)$ satisfy (3.40). We compute the asymptotic form of $\hat{\xi}_i(\varepsilon, \sigma)$ as $\varepsilon \downarrow 0$. Using (3.48) and (3.49), we have

$$\lim_{\varepsilon\downarrow 0}\langle \tilde{f}^\varepsilon_v \hat{\phi}^\varepsilon_{ij}, \tilde{\theta}^\varepsilon_j \rangle = \gamma^* \kappa^i_j \int_{-\infty}^{+\infty} f_v(\tilde{u}^*, v^*)\frac{d}{dy}\tilde{u}^* dy$$
$$\times \int_{-\ell+x^*_j}^{x^*_j} g(U^*_n, V^*_n)dx. \tag{3.51}$$

Note that $\tilde{u}^*(y)$ is strictly monotone increasing (resp. decreasing) depending on the quantity (3.48) being negative (resp. positive). For the increasing case, we have

$$\int_{-\infty}^{+\infty} f_v(\tilde{u}^*, v^*)\frac{d}{dy}\tilde{u}^* dy = \frac{d}{dv}\int_{h_-(v^*)}^{h_+(v^*)} f(s, v^*)ds$$
$$= \frac{dJ}{dv}(v^*) < 0 \qquad (\text{see (A.2)}). \tag{3.52}$$

For the decreasing case, the sign becomes opposite. On the other hand, we have from Remark 3.7 and (3.42),

$$\lim_{\varepsilon\downarrow 0}\langle \hat{\phi}^\varepsilon_{ij}, \tilde{u}^\varepsilon_y \rangle = \kappa^i_j \frac{(c^*)^2}{\gamma^*}. \tag{3.53}$$

Substituting (3.51), (3.52) and (3.53) into (3.50b), we finally obtain for the increasing case

$$\hat{\xi}^*_i \equiv \lim_{\varepsilon\downarrow 0}\hat{\xi}_i(\varepsilon, \sigma) = -\left(\frac{\gamma^*}{c^*}\right)^2 \frac{dJ}{dv}(v^*)\int_{-\ell+x^*_j}^{x^*_j} g(U^*_n, V^*_n)dx. \tag{3.54}$$

For the decreasing case, the integral part should be replaced by $\displaystyle\int_{-r+x^*_j}^{x^*_j} g(U^*_n, V^*_n)dx$

which is equal to $-\displaystyle\int_{-\ell+x^*_j}^{x^*_j} g(U^*_n, V^*_n)dx$. However, the sign of $\dfrac{dJ}{dv}(v^*)$ (see (3.52)) also

changes at the same time, hence $\hat{\xi}^*_i$ is equal to (3.54) in either case. Recalling Proposition

3.8, we see that $R^n(U^{*,\sigma/n^2}) = U_n^*$ and $R^n(V^{*,\sigma/n^2}) = V_n^*$, and hence we have

$$\int_{-\ell+x_j^*}^{x_j^*} g(U_n^*, V_n^*)dx = \frac{1}{n}\int_0^{x_j^*(\sigma/n^2)} g(U^{*,\sigma/n^2}, V^{*,\sigma/n^2})dx,$$

which completes the proof of (3.39). The final part of Lemma 3.15 was already shown in Lemma 3.10. $\square$

3.3. Derivation of the SLEP System

The most delicate and crucial part of the linearized spectral analysis for layered solutions is to find all critical eigenvalues of (3.12) and clarify their asymptotic behaviors as $\varepsilon \downarrow 0$. The SLEP method gives us a unified tool to deal with this problem. The basic idea of it is to find a *nice scaling* which blows up the degenerate situation of $\mathcal{L}^{\varepsilon,\sigma}$ (see (3.12)) as $\varepsilon \downarrow 0$. It turns out that the study of asymptotic behaviors of critical eigenvalues is reduced to solving a linear eigenvalue problem of $n \times n$ *symmetric* matrix called the *SLEP system*, although (3.12) is not self-adjoint. The aim of this section is to show how the linearized problem (3.12) is reduced to the SLEP system with respect to the *scaled* critical eigenvalues. Without loss of generality, we can restrict the region of λ to Λ_1 defined by

$$\Lambda_1 \equiv \{\lambda \mid \mathrm{Re}\lambda > -\mu_1 > \max(-\Delta^*, -\mu)\} \tag{3.55}$$

for some fixed $\mu_1 > 0$, where $-\Delta^*$ and $-\mu$ are negative constants appeared in Lemma 3.15 and (3.17). First note the following lemma.

Lemma 3.17. *The first n eigenvalues $\{\zeta_i^{\varepsilon,\sigma}\}_{i=0}^{n-1}$ of Sturm-Liouville operator $L^{\varepsilon,\sigma}$ do not belong to the spectra of $\mathcal{L}^{\varepsilon,\sigma}$ for small ε.*

Proof. See Appendix B. –

Remark 3.18. *Lemma 3.17 combined with Lemma 3.15 implies that, if $\lambda \in \Lambda_1$ is an eigenvalue of $\mathcal{L}^{\varepsilon,\sigma}$, the resolvent $(L^{\varepsilon,\sigma} - \lambda)^{-1}$ exists for all small ε.*

Solving the first equation of (3.12) with respect to w and substituting it into the second equation after expanding it by using the complete orthonormal set of $L^{\varepsilon,\sigma}$, we have the equivalent eigenvalue problem containing only z:

$$\frac{1}{\sigma}\frac{d^2}{dx^2}z + \sum_{i=0}^{n-1}\frac{(-f_v^{\varepsilon,\sigma}z, \phi_i^{\varepsilon,\sigma})}{\zeta_i^{\varepsilon,\sigma}-\lambda}g_u^{\varepsilon,\sigma}\phi_i^{\varepsilon,\sigma} + g_u^{\varepsilon,\sigma}(L^{\varepsilon,\sigma}-\lambda)^\dagger(-f_v^{\varepsilon,\sigma}z) + g_v^{\varepsilon,\sigma}z = \lambda z \tag{3.56}$$
$$\lambda \in \Lambda_1,$$

where the *reduced resolvent* $(L^{\varepsilon,\sigma} - \lambda)^\dagger$ is defined by

$$(L^{\varepsilon,\sigma} - \lambda)^\dagger(\cdot) \equiv \sum_{i\geq n}\frac{\langle \cdot, \phi_i^{\varepsilon,\sigma}\rangle}{\zeta_i^{\varepsilon,\sigma} - \lambda}\phi_i^{\varepsilon,\sigma} : L^2(I) \to L^2(I) \cap \{\phi_0^{\varepsilon,\sigma}, \cdot, \phi_{n-1}^{\varepsilon,\sigma}\}^\perp. \tag{3.57}$$

It is clear from (3.57) that $(L^{\varepsilon,\sigma} - \lambda)^\dagger$ is uniformly L^2-bounded operator, more precisely, we have

$$\| (L^{\varepsilon,\sigma} - \lambda)^\dagger \| \leq \frac{1}{|\zeta_n^{\varepsilon,\sigma} - \lambda|} \qquad (3.58)$$

for $\lambda \in \Lambda_1$. Note that the denominator $|\zeta_n^{\varepsilon,\sigma} - \lambda|$ is strictly bounded away from zero (see Lemma 3.15 and (3.55)). On the other hand, recalling Lemma 3.15, all the denominators of the second term of (3.56) must go to zero as $\varepsilon \downarrow 0$ if λ is a critical eigenvalue of (3.12). Hence more careful treatment is needed to handle this term. Let us begin with the study of the reduced resolvent.

Lemma 3.19 (Asymptotic Limit of $(L^{\varepsilon,\sigma} - \lambda)^\dagger$). $(L^{\varepsilon,\sigma} - \lambda)^\dagger$ is a uniform L^2-bounded operator with respect to ε and becomes a multiplication operator in the limit of $\varepsilon \downarrow 0$. Namely,

$$\lim_{\varepsilon \downarrow 0}(L^{\varepsilon,\sigma} - \lambda)^\dagger (F^{\varepsilon,\sigma}h) = \frac{F^{*,\sigma}h}{f_u^{*,\sigma} - \lambda}$$

in strongly L^2-sense for any $h \in L^2(I) \cap L^\infty(I)$, smooth function $F(u, v)$, and $\lambda \in \Lambda_1$, where $F^{\varepsilon,\sigma} = F(D^n(\varepsilon, \sigma))$ and $F^{*,\sigma} = F(U_n^{*,\sigma}(x), V_n^{*,\sigma}(x))$ (i.e., evaluated at the reduced solution of $D^n(\varepsilon, \sigma)$). Moreover, if h belongs to $H^1(I)$, the above convergence is uniform on any bounded set in $H^1(I)$.

Proof. Let $S_\delta(x)$ be a smooth cut-off function defined on I satisfying

$$S_\delta(x) = \begin{cases} 1 & \text{if } |x| \geq \delta/2 \\ 0 & \text{if } |x| \leq \delta/4 \end{cases}$$

with

$$0 \leq S_\delta(x) \leq 1, \quad \sup_{x \in I}\left|\frac{d^i}{dx^i}S_\delta(x)\right| \leq M_\delta \quad (i = 1, 2),$$

where M_δ is a positive constant which tends to $+\infty$ as $\delta \downarrow 0$. We assume for simplicity that $F^{\varepsilon,\sigma} \equiv 1, \lambda = 0$, and h is smooth. It is not difficult to extend it for the general case by using approximation and density argument.

Define S_δ^i by $S_\delta^i \equiv S_\delta(x - x_i^*)$, then

$$h_\delta \equiv S_\delta^1 \cdot S_\delta^2 \cdots S_\delta^n \cdot h$$

is a punctured function of h at layer positions, i.e., $h_\delta \equiv 0$ in a small neighbourhood of $x_i^*(i = 1, \cdots, n)$. In view of (3.58), we see that it suffices to prove the lemma for h_δ since h_δ approaches h in L^2-sense as $\delta \downarrow 0$. In order to show that $h_\delta/f_u^{*,\sigma}$, which we denote by u_δ, is equal to $\lim_{\varepsilon\downarrow 0}(L^\varepsilon)^\dagger h_\delta$, we first apply $L^{\varepsilon,\sigma}$ to u_δ:

$$
\begin{aligned}
L^{\varepsilon,\sigma}u_\delta &= \varepsilon^2\frac{d^2}{dx^2}\left(\frac{h_\delta}{f_u^{*,\sigma}}\right)+f_u^{\varepsilon,\sigma}\left(\frac{h_\delta}{f_u^{*,\sigma}}\right) \\
&= h_\delta + \frac{(f_u^{\varepsilon,\sigma}-f_u^{*,\sigma})}{f_u^{*,\sigma}}h_\delta + \varepsilon^2 H_\delta,
\end{aligned}
\tag{3.59}
$$

where $H_\delta \equiv \dfrac{d^2}{dx^2}\left(\dfrac{h_\delta}{f_u^{*,\sigma}}\right)$. Note that H_δ is well-defined since h_δ becomes zero in a neighbourhood of each x_i^*. Applying $(L^{\varepsilon,\sigma})^\dagger$ to both sides of (3.59), we have

$$
u_\delta - \sum_{i=0}^{n-1}\langle u_\delta, \phi_i^{\varepsilon,\sigma}\rangle\phi_i^{\varepsilon,\sigma} = (L^{\varepsilon,\sigma})^\dagger h_\delta + (L^{\varepsilon,\sigma})^\dagger\left\{\frac{(f_u^{\varepsilon,\sigma}-f_u^{*,\sigma})}{f_u^{*,\sigma}}h_\delta + \varepsilon^2 H_\delta\right\}.
\tag{3.60}
$$

Recalling Corollary 3.14 and $u_\delta \equiv 0$ in a neighbourhood of each x_i^*, the second term on the left-hand side tends to zero in L^2-sense as $\varepsilon\downarrow 0$. Since $f_u^{\varepsilon,\sigma} - f_u^{*,\sigma}$ goes to zero in L^2-sense as $\varepsilon\downarrow 0$ and H_δ remains bounded in L^2-sense, the second term of the right-hand side of (3.60) also tends to zero when $\varepsilon\downarrow 0$. Hence we obtain

$$
u_\delta = \lim_{\varepsilon\downarrow 0}(L^{\varepsilon,\sigma})^\dagger h_\delta.
$$

The final part of the lemma can be proved in the same way as in the proof of Lemma 2.2 in [44], so we omit the details.

It is convenient to classify the spectrum of (3.12) into two classes; the *critical* eigenvalues which tend to zero as $\varepsilon\downarrow 0$, and *noncritical* eigenvalues which are bounded away from the imaginary axis for small ε. Making use of Lemma 3.19, we can show that noncritical eigenvalues are *not* dangerous to the stability of $D^n(\varepsilon,\sigma)$, namely, they have strictly negative real parts independent of ε.

Proposition 3.20 (A Priori Bound for Noncritical Eigenvalues). *Let B_δ be a closed ball with center at the origin and radius δ in the complex plane $\mathbf{C}$. Suppose that λ is an arbitrary noncritical eigenvalue of (3.12) which stays outside of B_δ for all small ε. Then there exist positive constants μ^* and ε_δ such that*

$$
Re\lambda < -\mu^* < 0 \quad for \quad 0 < \varepsilon < \varepsilon_\delta,
\tag{3.61}
$$

where μ^* does not depend on δ and ε.

Proof. There is no essential difference in proofs between mono-layer and multi-layer cases. So we delegate it to that of Proposition 2.1 in [44].

Now we can concentrate on the behavior of critical eigenvalues. Let $\lambda = \lambda(\varepsilon)$ be an arbitrary critical eigenvalue of (3.12), and assume that λ varies in the ball $B_\delta = \{\lambda\,|\,|\lambda| < \delta\}$ for some $\delta > 0$. We shall specify the size of δ later. In view of Corollary 3.14, Lemma 3.15, and $z \in H_N^1(I)$, we see that both the denominator and the numerator of the second term on the left-hand side of (3.56) tend to zero as $\varepsilon\downarrow 0$. Here the scaling technique comes up to convert it into more tractable and nondegenerate form. Before that, we first rewrite (3.56) in the form of a *finite dimensional eigenvalue problem* by using the following operator $K^{\varepsilon,\sigma,\lambda}$.

Lemma 3.21 (Operator $K^{\varepsilon,\sigma,\lambda}$). *Let $\hat{B}^{\varepsilon,\sigma,\lambda}$ be a bilinear form defined by*

$$\hat{B}^{\varepsilon,\sigma,\lambda}(z^1, z^2) = \frac{1}{\sigma}\langle z_x^1, z_x^2 \rangle - \langle \{g_u^{\varepsilon,\sigma}(L^{\varepsilon,\sigma} - \lambda)^{\dagger}(-f_v^{\varepsilon,\sigma}\cdot) + g_v^{\varepsilon,\sigma} - \lambda\}z^1, z^2 \rangle$$

$$\text{for} \quad z^i \in H_N^1(I) \ (i = 1, 2).$$

Then, for a given $h \in H^{-1}(I)$, the equation for $z \in H_N^1(I)$

$$\hat{B}^{\varepsilon,\sigma,\lambda}(z, \psi) = \langle h, \psi \rangle \quad \text{for any} \quad \psi \in H_N^1(I)$$

has a unique solution $z = z(h)$ for small ε (including $\varepsilon = 0$) and $\lambda \in B_\delta$. Define the mapping $K^{\varepsilon,\sigma,\lambda}$ by

$$K^{\varepsilon,\sigma,\lambda}h = z(h); \ H^{-1}(I) \longrightarrow H_N^1(I).$$

$K^{\varepsilon,\sigma,\lambda}$ *is a bounded operator from $H^{-1}(I)$ to $H_N^1(I)$, and depends continuously on (ε, σ) and analytically on λ in operator norm sense, respectively. The limiting form $K^{*,\sigma,0} \equiv \lim_{\varepsilon \downarrow 0} K^{\varepsilon,\sigma,0}$ is given by (3.68) and (3.69).*

Proof. See Lemma 3.1 of [44].

Applying $K^{\varepsilon,\sigma,\lambda}$ to (3.56), we have

$$z = \sum_{i=0}^{n-1} \frac{\langle -f_v^{\varepsilon,\sigma}z, \phi_i^{\varepsilon,\sigma} \rangle}{\zeta_i^{\varepsilon,\sigma} - \lambda} K^{\varepsilon,\sigma,\lambda}(g_u^{\varepsilon,\sigma}\phi_i^{\varepsilon,\sigma}). \tag{3.62}$$

This shows that z is a linear combination of $K^{\varepsilon,\sigma,\lambda}(g_u^{\varepsilon,\sigma}\phi_i^{\varepsilon,\sigma})$ $(i = 0, \cdots, n-1)$ yielding

$$z = \sum_{i=0}^{n-1} \alpha_i K^{\varepsilon,\sigma,\lambda}(g_u^{\varepsilon,\sigma}\phi_i^{\varepsilon,\sigma}), \tag{3.63}$$

where $\mathbf{a} = (\alpha_0, \cdots, \alpha_{n-1})$ is a real vector. Note that $K^{\varepsilon,\sigma,\lambda}(g_u^{\varepsilon,\sigma}\phi_i^{\varepsilon,\sigma})$ $(i = 0, \cdots, n-1)$ are linearly independent since $\{\phi_i^{\varepsilon,\sigma}\}_{i=0}^{n-1}$ are linearly independent. Substituting (3.63) into (3.62), we obtain an *n-dimensional* matrix eigenvalue problem:

$$M^{\varepsilon,\sigma}\mathbf{a} = \begin{pmatrix} (\zeta_0^{\varepsilon,\sigma} - \lambda)\alpha_0 \\ \vdots \\ (\zeta_{n-1}^{\varepsilon,\sigma} - \lambda)\alpha_{n-1} \end{pmatrix}, \tag{3.64}$$

where $M^{\varepsilon,\sigma} = \{\langle -f_v^{\varepsilon,\sigma}\phi_i^{\varepsilon,\sigma}, K^{\varepsilon,\sigma,\lambda}(g_u^{\varepsilon,\sigma}\phi_j^{\varepsilon,\sigma})\rangle\}_{i,j=0}^{n-1}$. Note that $M^{\varepsilon,\sigma}$ also depends on λ through $K^{\varepsilon,\sigma,\lambda}$. This problem is highly degenerated, since all the elements of $M^{\varepsilon,\sigma}$ and $\zeta_i^{\varepsilon,\sigma} - \lambda$ $(i = 0, \cdots, n-1)$ tend to zero as $\varepsilon \downarrow 0$ (recall Corollary 3.14, Lemma 3.15, Lemma 3.21, and that λ is a critical eigenvalue). The following characterization of the asymptotic form of $\phi_i^{\varepsilon,\sigma}$ by $\sqrt{\varepsilon}$-scaling plays a key role to unfold this degenerate situation.

Lemma 3.22 (Asymptotic Form of $\phi_i^{\varepsilon,\sigma}/\sqrt{\varepsilon}$). *Let $\{\phi_i^{\varepsilon_m,\sigma}\}_{m=1}^{\infty}$ be an arbitrary convergent sequence in the sense of Lemma 3.12 on each stretched subinterval $\tilde{I}_j (j = 1, \cdots, n)$ for $i \in (0, \cdots, n-1)$. Then it holds that*

$$\lim_{m\uparrow\infty} -f_v^{\varepsilon_m,\sigma} \frac{\phi_i^{\varepsilon_m,\sigma}}{\sqrt{\varepsilon}} = c_1^* \Delta_i \equiv c_1^* \sum_{j=1}^{n} \kappa_j^i \delta(x - x_j^*(\sigma)), \tag{3.65a}$$

$$\lim_{m\uparrow\infty} g_u^{\varepsilon_m,\sigma} \frac{\phi_i^{\varepsilon_m,\sigma}}{\sqrt{\varepsilon}} = c_2^* \Delta_i \equiv c_2^* \sum_{j=1}^{n} \kappa_j^i \delta(x - x_j^*(\sigma)) \tag{3.65b}$$

both in $H^{-1}(I)$ -sense, where $c_1^* \equiv -\gamma^* \frac{dJ}{dv}(v^*)$, $c_2^* \equiv \gamma^* \{g(h_+(v^*), v^*) - g(h_-(v^*), v^*)\}$, and $\delta(x - x_j^*(\sigma))$ denotes the Dirac's δ-function at $x = x_j^*(\sigma)$. The vectors $\mathbf{k}^i = (\kappa_1^i, \cdots, \kappa_n^i)$ $(i = 0, \cdots, n-1)$ satisfy the orthogonal relation (3.35).

Proof. Recallng (3.26) and Lemma 3.11, we see that this is essentially a restatement of Lemma 3.12 in the original x-coordinate. The only difference is the coefficients c_i^* $(i = 1, 2)$ which appear due to the existence of $-f_v^{\varepsilon_m,\sigma}$ and $g_u^{\varepsilon_m,\sigma}$. To show (3.65a), it suffices to compute the integral

$$\lim_{\varepsilon\downarrow0} \int_{\tilde{I}_j} \sqrt{\varepsilon}(-\tilde{f}_v^{\varepsilon_m,\sigma}) \tilde{\phi}_{ij}^{\varepsilon_m,\sigma} dy = \int_{-\infty}^{\infty} -\tilde{f}_v^{\varepsilon_m,\sigma} \kappa_j^i \hat{\phi}_L^* dy$$

$$= \int_{-\infty}^{\infty} -\tilde{f}_v^{\varepsilon_m,\sigma} \kappa_j^i \gamma^* \frac{d}{dy} \tilde{u}^* dy$$

$$= -\gamma \kappa_j^i \int_{h_-(v^*)}^{h_+(v^*)} f_v(s, v^*) ds$$

$$= -\kappa_j^i \gamma \frac{d}{dv} J(v^*)$$

$$= c_1^* \kappa_j^i,$$

which implies (3.65a). Similar computation leads to (3.65b).

Hereafter, we fix a convergent subsequence $\{\phi_i^{\varepsilon_m,\sigma}/\sqrt{\varepsilon}\}_{m=1}^{\infty}$, and for simplicity of notation, we simply write ε instead of ε_m keeping in mind that ε actually means a discrete parameter ε_m. In view of Lemma 3.15 and Lemma 3.22, we see that ε-scaling is the most suitable to blow up (3.64). In fact, dividing (3.64) by ε on both sides, we have

$$\tilde{M}^{\varepsilon,\sigma} \mathbf{a} = \begin{pmatrix} (\zeta_0^{\varepsilon,\sigma}/\varepsilon - \lambda/\varepsilon)\alpha_0 \\ \vdots \\ (\zeta_{n-1}^{\varepsilon,\sigma}/\varepsilon - \lambda/\varepsilon)\alpha_{n-1} \end{pmatrix}, \tag{3.66}$$

where $\tilde{M}^{\varepsilon,\sigma} = \{(-f_v^{\varepsilon,\sigma}\phi_i^{\varepsilon,\sigma}/\sqrt{\varepsilon}, K^{\varepsilon,\sigma,\lambda}(g_u^{\varepsilon,\sigma}\phi_j^{\varepsilon,\sigma}/\sqrt{\varepsilon}))\}_{i,j=0}^{n-1}$. The problem (3.66) is non-degenerate and well-defined continuously up to $\varepsilon = 0$. In fact, in the limit of $\varepsilon \downarrow 0$, we see from Lemmas 3.15, 3.21, and 3.22 that (3.66) becomes

$$\{\tilde{M}^{*,\sigma} + (\tau^{*,\sigma} - \sigma\zeta^{*,\sigma})I\}\mathbf{a}^* = 0 \qquad \text{(SLEP system)} \qquad (3.67)$$

where $\tilde{M}^{*,\sigma} \equiv \lim_{\varepsilon\downarrow 0} \tilde{M}^{\varepsilon,\sigma} = \{c_1^* c_2^* \langle \Delta_i, K^{*,\sigma}\Delta_j \rangle\}_{i,j=0}^{n-1}$, $K^{*,\sigma} \equiv K^{0,\sigma,0}$, $\tau^{*,\sigma} \equiv \lim_{\varepsilon\downarrow 0} \lambda(\varepsilon)/\varepsilon$, $\sigma\zeta^{*,\sigma} \equiv \lim_{\varepsilon\downarrow 0} \zeta_i^{\varepsilon,\sigma}/\varepsilon$ $(i = 0, \cdots, n - 1)$, and $\mathbf{a}^* = \lim_{\varepsilon\downarrow 0} \mathbf{a}$. Here we use the fact that the critical eigenvalue $\lambda = \lambda(\varepsilon)$ can be written in the following form:

Lemma 3.23. *Any critical eigenvalue λ must have the form*

$$\lambda = \varepsilon\tau(\varepsilon, \sigma),$$

where τ is a bounded function up to $\varepsilon = 0$. The problem (3.66) depends on τ smoothly.

Proof. Suppose that there is a critical eigenvalue $\lambda_c(\varepsilon)$ which tends to zero strictly slower than $O(\varepsilon)$. Then the associated eigenvalue problem (3.66) must have an arbitrary large (in modulus) eigenvalue when $\varepsilon \downarrow 0$. However this is not possible since both $\tilde{M}^{\varepsilon,\sigma}$ and $\zeta^{\varepsilon,\sigma}$ are uniformly bounded and have definite limits as $\varepsilon \downarrow 0$. This also implies the boundedness of τ up to $\varepsilon = 0$. The last part is clear from Lemma 3.21.

The limiting problem (3.67) is called the *SLEP system* of (3.12) with respect to the scaled eigenvalues $\tau^{*,\sigma}$. From now on, we focus on the SLEP system (3.67), since all information on the asymptotic behaviors of critical eigenvalues for $\varepsilon > 0$ can be derived from (3.67). However the only defect of (3.67) is that it does not look free from the choice of the subsequence, i.e., $\tilde{M}^{*,\sigma}$ may depend on Δ_i. In order to get rid of this ambiguity, we shall apply a change of bases to $\tilde{M}^{*,\sigma}$ which depends on $\{\Delta_i\}_{i=0}^{n-1}$. It turns out that the resulting matrix denoted by $\tilde{G}_N$ is *independent* of the choice of the subsequence and has n real distinct eigenvalues, which leads to our main result (Theorem 3.25). For this purpose, we introduce the Green function $G_N = G_N(x, y; \sigma)$ associated with the operator $K^{*,\sigma,0}$ (see Lemma 3.21) defined as follows:

$$K^{*,\sigma,0}\phi = \langle G_N(x, y; \sigma), \phi \rangle, \quad \text{for any} \quad \phi \in H^{-1}(I). \qquad (3.68)$$

More explicitly,

$$G_N(x, y; \sigma) = -\frac{\sigma}{W(h, k)} \times \begin{cases} h(x)k(y), & 0 \le x \le y \le 1, \\ h(y)k(x), & 0 \le y \le x \le 1, \end{cases} \qquad (3.69a)$$

where h and k satisfy the equation

$$\left(-\frac{1}{\sigma}\frac{d^2}{dx^2} - \frac{\det^{*,\sigma}}{f_u^{*,\sigma}}\right)\phi = 0, \qquad \phi \in H^2(I) \qquad (3.69b)$$

with the boundary conditions

$$h(0) = 1, h'(0) = 0 \quad \text{and} \quad k(1) = 1, k'(1) = 0, \qquad (3.69c)$$

where $\det^{*,\sigma} \equiv f_u^{*,\sigma} g_v^{*,\sigma} - f_v^{*,\sigma} g_u^{*,\sigma} > 0$ (see (A.4)) and $W(h, k)$ denotes the Wronskian of h and k. Note that h (resp. k) is strictly positive and increasing (resp. decreasing), respectively, since $-\det^{*,\sigma} / f_u^{*,\sigma}$ is strictly positive from (A.3) and (A.4). It follows from (3.68) that

$$\langle \delta(x - x_i^*(\sigma)), K^{*,\sigma,0}\delta(x - x_j^*(\sigma)) \rangle = G_N(x_i^*(\sigma), x_j^*(\sigma); \sigma).$$

Therefore, recalling that $\Delta_i = \displaystyle\sum_{j=1}^{n} \kappa_j^i \delta(x - x_j^*(\sigma))$, we have

$$\langle \Delta_i, K^{*,\sigma,0}\Delta_j \rangle = {}^t\mathbf{k}^i \mathbf{G}_N \mathbf{k}^j, \tag{3.70}$$

where $\mathbf{G}_N$ is an $n \times n$ symmetric matrix with positive components defined by

$$\mathbf{G}_N \equiv \{G_N(x_i^*(\sigma), x_j^*(\sigma); \sigma)\}_{i,j=1}^{n}. \tag{3.71}$$

Let us define matrix P by

$$P \equiv c^*(\mathbf{k}^0, \mathbf{k}^1, \cdots, \mathbf{k}^{n-1}) \tag{3.72}$$

which becomes orthogonal from (3.35). Then the matrix $\tilde{M}^{*,\sigma}$ in (3.67) can be rewritten as

$$\begin{aligned}
\tilde{M}^{*,\sigma} &= \{c_1^* c_2^* {}^t\mathbf{k}^i \mathbf{G}\mathbf{k}^j\}_{i,j=0}^{n-1} \\[2mm]
&= \frac{c_1^* c_2^*}{(c^*)^2} {}^t P \mathbf{G}_N P \tag{3.73} \\[2mm]
&= {}^t P \tilde{\mathbf{G}}_N P,
\end{aligned}$$

where

$$\tilde{\mathbf{G}}_N \equiv \frac{c_1^* c_2^*}{(c^*)^2} \mathbf{G}_N. \qquad \text{(SLEP matrix)} \tag{3.74}$$

This shows that $\tilde{M}^{*,\sigma}$ is an orthogonal transformation of the real symmetric matrix $\tilde{\mathbf{G}}_N$ which clearly does *not* depend on the choice of the subsequence, and hence neither do the eigenvalues of $\tilde{M}^{*,\sigma}$. We reach the final form by multiplying both side of (3.67) by P:

$$\{\tilde{\mathbf{G}}_N + (\tau^{*,\sigma} - \sigma\hat{\xi}^{*,\sigma})I\}\mathbf{b}^* = 0, \tag{3.75}$$

where $\mathbf{b}^* \equiv Pa^*$. This is, what we call, *the normal SLEP system* of (3.12). Since the normal SLEP system (3.75) is the common limit of all subsequences, the eigenvalues of (3.66) (without taking subsequence) converge to those of (3.75) when $\varepsilon \downarrow 0$. Moreover, since $\tilde{\mathbf{G}}_N$ has n real distinct eigenvalues (see Theorem 3.24), the associated eigenfunctions as well as eigenvalues of (3.66) are uniquely determined as continuous functions of ε by usual *regular* perturbation. What we have to do is to show that $\tilde{\mathbf{G}}_N$ has n distinct real eigenvalues and determine their signs. Especially, we are interested in the minimum eigenvalue of $\tilde{\mathbf{G}}_N$, since it determines the maximum value of the scaled critical eigenvalues $\tau^{*,\sigma}$. If the maximum value of $\tau^{*,\sigma}$ is negative (resp. positive), the normal n-layered solution becomes asymptotically stable (resp. unstable). We have the following result for the eigenvalue problem (3.75), the proof of which is delegated to Section 3.4.

Theorem 3.24 (Eigenvalues of the SLEP System). *The set of eigenvalues of $\tilde{\mathbf{G}}_N$*

$$\tilde{\mathbf{G}}_N \theta = \gamma \theta \tag{3.76}$$

consists of n real distinct positive eigenvalues

$$0 < \gamma_{n-1} < \gamma_{n-2} < \cdots < \gamma_0. \tag{3.77}$$

Namely, in term of $\tau^{*,\sigma}$, (3.75) has n distinct real eigenvalues

$$\tau_0^{*,\sigma} < \tau_1^{*,\sigma} < \cdots < \tau_{n-1}^{*,\sigma}, \tag{3.78}$$

where $\tau_i^{*,\sigma} = \sigma \hat{\zeta}^{*,\sigma} - \gamma_i \ (0 \leq i \leq n - 1)$. Moreover, it holds that

$$\tau_{n-1}^{*,\sigma} < 0. \tag{3.79}$$

Theorem 3.24 leads us to the following main result by a regular perturbation. Asymptotic forms of critical eigenfunctions are also given in the next theorem.

Theorem 3.25. *There is a positive constant $\hat{\varepsilon}$ such that the critical eigenvalues of $\mathcal{L}^{\varepsilon,\sigma}$ consist of n real distinct eigenvalues $\lambda_0^c(\varepsilon), \cdots, \lambda_{n-1}^c(\varepsilon)$ which are simple and continuous for $0 \leq \varepsilon < \hat{\varepsilon}$ satisfing the asymptotic relations (see Figure 3.3)*

$$\lambda_k^c(\varepsilon) \simeq \tau_k^{*,\sigma} \varepsilon, \qquad k = 0, \cdots, n - 1$$

as $\varepsilon \downarrow 0$, where $\tau_k^{*,\sigma} \ (k = 0, \cdots, n - 1)$ are given in Theorem 3.24. The associated critical eigenfunction $\Phi^k(\varepsilon) = {}^t(w^k(\varepsilon), z^k(\varepsilon))$ has the following asymptotic form

$$\lim_{\varepsilon \downarrow 0} \Phi^k(\varepsilon) \equiv \Phi^{k^*} = \begin{pmatrix} w^{k^*} \\ z^{k^*} \end{pmatrix} = \begin{pmatrix} \displaystyle\sum_{j=1}^{n} q_j^k \delta_{x_j} - \frac{f_v^{*,\sigma}}{f_u^{*,\sigma}} c_2^* \sum_{j=1}^{n} q_j^k K^{*,\sigma,0} \delta_{x_j} \\ \\ \displaystyle c_2^* \sum_{j=1}^{n} q_j^k K^{*,\sigma,0} \delta_{x_j} \end{pmatrix}, \tag{3.80}$$

where $\mathbf{q}^k = (q_1^k, \cdots, q_n^k) \ (\|q^k\| = 1)$ is an eigenvector of $\tilde{\mathbf{G}}_N$ for γ_k, and we use the simple notation δ_{x_j} instead of $\delta(x - x_j^*(\sigma))$. The rest of the eigenvalues, i.e., noncritical ones, have strictly negative real parts uniformly for $0 < \varepsilon < \hat{\varepsilon}$.

Remark 3.26. *Note that asymptotic forms (3.80) are free from the chosen subsequence in Lemma 3.22.*

Proof. Since we know from Theorem 3.24 that $\{\tau_i^{*,\sigma}\}_{i=0}^{n-1}$ are real and distinct, the first part of the theorem is easily obtained by applying the implicit function theorem to the characteristic equation of (3.66) with replacing λ by $\varepsilon\tau$. In fact there exist ε^* and κ^* such that (3.66) has a unique ε-family of solutions $\{\tau_k(\varepsilon, \sigma)\}_{k=0}^{n-1}$ for $0 < \varepsilon < \varepsilon^*$ and $|\tau - \tau_k^{*,\sigma}| < \kappa^*$ satisfying $\lim_{\varepsilon \downarrow 0} \tau_k(\varepsilon, \sigma) = \tau_k^{*,\sigma} \ (k = 0, \cdots, n - 1)$. Conversely we can find $\varepsilon_{\kappa_0} \ (\leq \varepsilon^*)$ and $\kappa_0 \ (> |\tau_0^{*,\sigma}|)$ such that any solution of (3.66) with $0 < \varepsilon < \varepsilon_{\kappa_0}$ and $|\tau| < \kappa_0$ must coincide with one of $\{\tau_k(\varepsilon, \sigma)\}_{k=0}^{n-1}$. Let $\hat{\delta} = \varepsilon_{\kappa_0} \kappa_0$. Then, taking

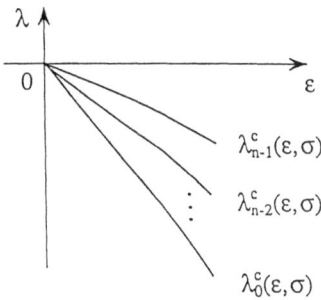

Figure 3.3.

$\delta = \hat{\delta}$ in Proposition 3.20, we see that noncritical eigenvalues of (3.12) satisfy (3.61) for $0 < \varepsilon < \hat{\varepsilon}$ where $\hat{\varepsilon} = \min\{\varepsilon_{\kappa_0}, \varepsilon_{\hat{\delta}}\}$. Simplicity of each critical eigenvalue can be proved in a parallel way as in the proof of Theorem 3.1 of [44]. The only remaining thing is to show (3.80). Denote by $Q = (\mathbf{q}^0, \cdots, \mathbf{q}^{n-1})$ the orthogonal matrix which diagonalizes $\tilde{G}_N$, i.e.,

$${}^t Q \tilde{G}_N Q = D_n, \tag{3.81}$$

where $D_n = diag(\gamma_0, \gamma_2, \cdots, \gamma_{n-1})$. Recalling that the matrix $\tilde{G}_N$ is represented with respect to the base $\{c_2^* K^{*,\sigma,0}(\delta_{x_i})\}_{i=1}^n$ (see (3.63), (3.72), (3.73)), it is clear that $z^{k^*} = c_2^* \sum_{j=1}^n q_j^k K^{*,\sigma,0} \delta_{x_j}$ is an eigenfunction associated with γ_k (and hence $\tau_k^{*,\sigma}$). As for the w-component, we have the expression for w by solving the first equation of (3.12)

$$w = \sum_{i=0}^{n-1} \frac{\langle z, -f_v^{\varepsilon,\sigma} \frac{\phi_i^{\varepsilon,\sigma}}{\sqrt{\varepsilon}} \rangle}{\zeta_i^{\varepsilon,\sigma} - \tau} \frac{\phi_i^{\varepsilon,\sigma}}{\sqrt{\varepsilon}} + (L^{\varepsilon,\sigma} - \lambda)^\dagger (-f_v^{\varepsilon,\sigma} z). \tag{3.82a}$$

Using Lemma 3.19 and 3.22, we see that w^{k^*} can be represented as

$$w^{k^*} = \sum_{i=0}^{n-1} \frac{c_1^* \langle z^{k^*}, \Delta_i \rangle}{\zeta_i^{*,\sigma} - \tau_k^{*,\sigma}} \Delta_i - \frac{f_v^{*,\sigma}}{f_u^{*,\sigma}} z^{k^*}. \tag{3.82b}$$

Substituting the above expression of z^{k^*} into (3.82b), we obtain

$$w^{k^*} = c_1^* c_2^* \sum_{i=0}^{n-1} \frac{\langle \sum_{j=0}^n q_j^k K^{*,\sigma,0} \delta_{x_j}, \sum_{\ell=0}^n \kappa_\ell^i \delta_{x_\ell} \rangle}{\zeta_i^{*,\sigma} - \tau_k^{*,\sigma}} \Delta_i$$

$$- c_2^* \frac{f_v^{*,\sigma}}{f_u^{*,\sigma}} \sum_{j=0}^n q_j^k K^{*,\sigma,0} \delta_{x_j}. \tag{3.83}$$

Since we want to have an asymptotic form independent of $\{\mathbf{k}^i\}_{i=0}^{n-1}$, we rewrite the first term on the right-hand side of (3.83). First note that $G_{j\ell} = \langle K^{*,\sigma,0}\delta_{x_j}, \delta_{x_\ell}\rangle$, we see from (3.81) that the numerator becomes

$$c_1^* c_2^* \sum_{j,\ell=1}^{n} G_{j\ell} q_j^k \kappa_\ell^i = (c^*)^2 \gamma_k \sum_{\ell=1}^{n} q_\ell^k \kappa_\ell^i.$$

Hence the first term on the right-hand side of (3.83) becomes

$$\frac{(c^*)^2 \gamma_k}{\hat{\zeta}_i^{*,\sigma} - \tau_k^{*,\sigma}} \sum_{i,j,\ell=1}^{n} q_\ell^k \kappa_\ell^i \kappa_j^i \delta_{x_j}.$$

Recalling that $P^t P = I$, where $P = c^*(\mathbf{k}^0, \cdots, \mathbf{k}^{n-1})$, this is equal to

$$\sum_{j=1}^{n} q_j^k \delta_{x_j}.$$

Here we use the relation $\hat{\zeta}^{*,\sigma} - \tau_k^{*,\sigma} = \gamma_k$. Combining the above results, we conclude that

$$w^{k*} = \sum_{j=1}^{n} q_j^k \delta_{x_j} - c_2^* \frac{f_v^{*,\sigma}}{f_u^{*,\sigma}} \sum_{j=0}^{n} q_j^k K^{*,\sigma,0} \delta_{x_j},$$

which completes the proof. $\square$

Combining Theorem 3.25 with Corollary 3.9, we obtain the following result which is equivalent to Main Theorem in Section 1.

Theorem 3.27. *For any fixed σ and positive integer n with $0 < \sigma < \sigma_0$, there exists $\varepsilon_n(\sigma) > 0$ such that the normal n-layered solution is asymptotically stable for $0 < \varepsilon < \varepsilon_n(\sigma)$, where $\varepsilon_n(\sigma)$ depends on n and σ with $\lim_{n\uparrow\infty} \varepsilon_n(\sigma) = 0$. Hence the number of asymptotically stable normal multi-layered solutions tends to infinity as $\varepsilon \downarrow 0$.*

3.4. Eigenvalue Problem for the SLEP Matrix

We shall prove Theorem 3.24 by sequence of lemmas. A key trick is to consider the eigenvalue problem of the *inverse* of $\tilde{\mathbf{G}}_N$, not $\tilde{\mathbf{G}}_N$ itself. The main feature of $\tilde{\mathbf{G}}_N^{-1}$ is that it is a tri-diagonal symmetric matrix (Lemma 3.28), which is well studied, and a general theory (Lemma 3.29) can be applied to $\tilde{\mathbf{G}}_N^{-1}$ to conclude Theorem 3.24 except (3.79). In order to show (3.79), which is crucial for the stability, we need to introduce the auxiliary matrix $\tilde{\mathbf{G}}_D$ which is similarly defined as $\tilde{\mathbf{G}}_N$ with replacing the Neumann boundary conditions by Dirichlet ones. Namely,

$$\tilde{\mathbf{G}}_D \equiv \frac{c_1^* c_2^*}{(c^*)^2} \mathbf{G}_D \equiv \frac{c_1^* c_2^*}{(c^*)^2} \{G_D(x_i^*(\sigma), x_j^*(\sigma); \sigma)\}_{i,j=1}^n, \tag{3.84}$$

where $G_D(x, y; \sigma)$ is the Green function of the operator (3.69b) under homogeneous Dirichlet boundary conditions. A key observation for $\tilde{\mathbf{G}}_D$ is that the minimum eigenvalue of it is equal to $\sigma \hat{\zeta}^{*,\sigma}$, or, in terms of $\tau^{*,\sigma}$, the maximum eigenvalue of (3.75) with replacing $\tilde{G}_N$ by $\tilde{G}_D$ is equal to zero (Lemma 3.31). This comes from the fact that the x-derivative of the normal n-layered solution is always an eigenfunction of (3.12) under Dirichlet boundary conditions associated with the *zero* eigenvalue, i.e.,

$$L^{\varepsilon,\sigma}u_x^{\varepsilon} + f_v^{\varepsilon,\sigma}v_x^{\varepsilon} = 0$$

$$M^{\varepsilon,\sigma}v_x^{\varepsilon} + g_u^{\varepsilon,\sigma}u_x^{\varepsilon} = 0 \tag{3.85}$$

$$u_x^{\varepsilon} = 0 = v_x^{\varepsilon} \qquad \text{on } \partial I.$$

A comparison between components of $\tilde{G}_N^{-1}$ and $\tilde{G}_D^{-1}$ (Lemma 3.32) leads to the conclusion (3.79). In view of (3.74) and (3.84), it suffices to compare G_N^{-1} with G_D^{-1}.

Lemma 3.28 (Inverse of the SLEP matrix). *The inverse of G_N exists and G_N^{-1} is a tri-diagonal real symmetric matrix such that*

(a) *All diagonal elements are equal except $(1, 1)$- and (n, n)-components.*
(b) *Every other off-diagonal elements are equal.*
(c) *All diagonal (resp. off-diagonal) elements are positive (resp. negative).*

More precisely we have

$$G_N^{-1} = \frac{W(h, k)}{\sigma} \begin{pmatrix} -\dfrac{h_2/h_1}{A_{12}} & \dfrac{1}{A_{12}} & & & & \\ & -\dfrac{A_{13}}{A_{12}A_{23}} & \dfrac{1}{A_{23}} & & & \\ & & \ddots & \ddots & & \\ & & & \ddots & \dfrac{1}{A_{12}} & \\ & & & & -\dfrac{h_2/h_1}{A_{12}} & \end{pmatrix} \quad , n : \text{even}, \quad (3.86)$$

$$G_N^{-1} = \frac{W(h, k)}{\sigma} \begin{pmatrix} -\dfrac{h_2/h_1}{A_{12}} & \dfrac{1}{A_{12}} & & & & \\ & -\dfrac{A_{13}}{A_{12}A_{23}} & \dfrac{1}{A_{23}} & & & \\ & & \ddots & \ddots & & \\ & & & \ddots & \dfrac{1}{A_{23}} & \\ & & & & -\dfrac{k_{n-1}/k_n}{A_{23}} & \end{pmatrix} \quad , n : \text{odd}, \quad (3.87)$$

where

$$A_{ij} \equiv \begin{vmatrix} h_i & k_i \\ h_j & k_j \end{vmatrix}$$

$$h_i \equiv h(x_i^*(\sigma)), \quad k_i \equiv k(x_i^*(\sigma)) \quad (see\ (3.69a)\ and\ (3.71)).$$

Exactly the same formulae hold for $\mathbf{G}_D^{-1}$ *with replacing* h_i, k_j *by* h_i^d, k_j^d. *Here* h_i^d, k_j^d *are the correspondents of the Green function for the Dirichlet boundary conditions.*

Proof. See Appendix C. —

The following result is basic for the study of the eigenvalue problem of tri-diagonal matrix. For the proof, see, for instance, Wilkinson [60; chapter 5].

Lemma 3.29 (Eigenvalues and Eigenfunctions of Tri-diagonal Symmetric Matrix). *Let* T *be a symmetric tri-diagonal matrix with non-zero off-diagonal elements of the form*

$$T = \begin{pmatrix} \alpha_1 & \beta_2 & & & & O \\ \beta_2 & \alpha_1 & \cdot & & & \\ & \cdot & \cdot & \cdot & & \\ & & \cdot & \cdot & \beta_n & \\ O & & & \cdot & \cdot & \\ & & & \beta_{n-1} & \alpha_n \end{pmatrix}, \quad \beta_i \neq 0.$$

Then, T *has* n *real, distinct, and simple eigenvalues* $\lambda_1 < \lambda_2 < \cdots < \lambda_n$. *The corresponding eigenvector* $\mathbf{x}^k =^t (x_1^k, \cdots, x_n^k)$ *to* λ_k *is expressed by*

$$x_1^k = 1, \quad x_r^k = (-1)^{r-1} p_{r-1}(\lambda_k)/\beta_2\beta_3 \cdots \beta_r \ (2 \leq r \leq n), \tag{3.88}$$

where $p_r(\lambda)$ *denotes the leading principal minor of order* r *of* $(T - \lambda I)$ *with* $p_0(\lambda) \equiv 1$. *Finally, the polynomials* $p_0(\lambda), p_1(\lambda), \cdots, p_n(\lambda)$ *satisfy the Sturm sequence property. Namely, let the quantities* $p_0(\mu), p_1(\mu), \cdots, p_n(\mu)$ *be evaluated for some value of* μ. *Then* $s(\mu)$, *the number of agreements in sign of consecutive members of this sequence, is equal to the number of eigenvalues of* T *which are strictly greater than* μ

Corollary 3.30 (Positivity of Eigenvalues of $\mathbf{G}_N$). *All eigenvalues of* $\mathbf{G}_N^{-1}$ *(and hence* $\mathbf{G}_N$ *also) are strictly positive. This is also true for* $\mathbf{G}_D^{-1}$.

Proof. It suffices to show that the minimum eigenvalue of $\mathbf{G}_N^{-1}$ is strictly positive. First, noting that all elements of $\mathbf{G}_N$ are non-negative, we see from the Perron-Frobenius Theorem (see Varga [59; chapter 2]) that the largest eigenvalue γ_{max} of it is real, simple, and positive with a positive eigenvector. Hence γ_{max}^{-1} must be one of the eigenvalues of $\mathbf{G}_N^{-1}$. On the other hand, the only eigenvalue of $\mathbf{G}_N^{-1}$ which has a positive eigenvector is the minimum one, say λ_1, because of (3.88) and the Sturm sequence property of Lemma 3.29. This implies $\lambda_1 = \gamma_{max}^{-1} > 0$ and completes the proof.

We need two more lemmas to prove (3.79) of Theorem 3.24. The first one is a restatement of the remark at the beginning of this subsection.

Lemma 3.31. *The minimum eigenvalue of* $\tilde{\mathbf{G}}_D$ *is equal to* $\sigma\hat{\zeta}^{*,\sigma}$. *In terms of* $\tau^{*,\sigma}$, *this is equivalent to say that the greatest eigenvalue of the problem (3.75) with replacing* $\tilde{\mathbf{G}}_N$ *by* $\tilde{\mathbf{G}}_D$ *is equal to zero.*

Proof. It is clear from (3.85) that $\sigma\hat{\zeta}^{*,\sigma}$ (i.e., $\tau = 0$) is one of the eigenvalues of $\tilde{\mathbf{G}}_D$. The only thing we have to prove is that it is the minimum eigenvalue of $\tilde{\mathbf{G}}_N$. To do this, note that Lemma 3.29 also holds for $\tilde{\mathbf{G}}_D^{-1}$, and therefore, the eigenvector associated with the largest eigenvalue has no agreement in sign of consecutive numbers of components. On the other hand, it is not difficult to see that the eigenvector of $\tilde{\mathbf{G}}_D$ associated with $\sigma\hat{\zeta}^{*,\sigma}$ also has the same property, since the original eigenvector $(u_x^\varepsilon, v_x^\varepsilon)$ of (3.85) can be generated by flipping the unit form on the first subinterval $(0, 1/n)$ by $(n-1)$-times in odd symmetric way. Hence the largest eigenvalue of $\tilde{\mathbf{G}}_D^{-1}$ must be $(\sigma\hat{\zeta}^{*,\sigma})^{-1}$, which implies from Corollary 3.30 that $\sigma\hat{\zeta}^{*,\sigma}$ is the minimum eigenvalue of $\tilde{\mathbf{G}}_D$.

A direct consequence of this lemma is that (3.79) is equivalent to

$$\text{Minimum eigenvalue of } \tilde{\mathbf{G}}_N > \text{Minimum eigenvalue of } \tilde{\mathbf{G}}_D = \sigma\hat{\zeta}^{*,\sigma}. \qquad (3.89)$$

In view of Corollary 3.30 this is equivalent to

$$\text{Maximum eigenvalue of } \tilde{\mathbf{G}}_N^{-1} < \text{Maximum eigenvalue of } \tilde{\mathbf{G}}_D^{-1}, \qquad (3.90)$$

which is more convenient for us. To show (3.90), we make a comparison between the elements of $\tilde{\mathbf{G}}_N^{-1}$ and $\tilde{\mathbf{G}}_D^{-1}$.

Lemma 3.32 (Comparison between $\tilde{\mathbf{G}}_N^{-1}$ and $\tilde{\mathbf{G}}_D^{-1}$). *It holds that*
(i) All the components of $\tilde{\mathbf{G}}_N^{-1}$ *and* $\tilde{\mathbf{G}}_D^{-1}$ *are equal except* $(1, 1)$ *and* (n, n) *components.*
(ii) For $(1, 1)$ *and* (n, n) *components, the following inequalities hold*

$$(\tilde{\mathbf{G}}_N^{-1})_{11} < (\tilde{\mathbf{G}}_D^{-1})_{11}$$

and

$$(\tilde{\mathbf{G}}_N^{-1})_{nn} < (\tilde{\mathbf{G}}_D^{-1})_{nn}.$$

Proof. See Appendix C.

Now we are ready to prove Theorem 3.24.

Proof of Theorem 3.24.
It is clear that Lemmas 3.28, 3.29, and Corollary 3.30 imply all the results of Theorem 3.24 except the stability inequality (3.79). Recalling the variational characterization of the principal eigenvalues of $\tilde{\mathbf{G}}_N^{-1}$ and $\tilde{\mathbf{G}}_D^{-1}$, the inequality (3.90) is a direct consequence of Lemma 3.32. Finally the inequality (3.90) combined with Lemma 3.31 leads us to the inequality (3.79).

4. Recovery Process of Stability

— From the Shadow System to the Full System —

In Section 3 we showed that, for a *fixed* $\sigma > 0$, all normal n-layered solutions are stable for small $\varepsilon > 0$. However their stabilities are subtle in the sense that they have n real negative critical eigenvalues which tend to zero with order ε as $\varepsilon \downarrow 0$. On the other hand, a single reaction diffusion equation has also similar normal n-layered solutions, but they are all unstable as was mentioned in Section 1.

A naive question is that "Can we somehow interpolate these two opposite results from spectral point of view?" A key ingredient for this is the shadow system (see (1.16) and Section 4.1) which is obtained as the limiting system when $\sigma \downarrow 0$, i.e., the diffusivity of the controller v is extremely high, and hence v is reduced to a constant function $v \equiv \xi(t)$ in spatial direction. The shadow system is an intermediate system between the full system and the single equation, and plays a pivotal role to answer the above question. Namely the spectra of the linearized problem (3.12) converge to those of the shadow system as $\sigma \downarrow 0$, and if the scalar controller ξ of the shadow system is fixed to be a constant with respect to time, it reduces to a single reaction diffusion equation.

In Section 4.1 we shall prove that the normal n-layered solutions to the shadow system have $(n-1)$ real *positive* eigenvalues which tend to zero exponentially as $\varepsilon \downarrow 0$, and the associated eigenfunctions belong to the D^n-symmetry breaking space (see Lemma 4.3). In Section 4.2 we shall show that the $(n-1)$ critical eigenvalues $\{\lambda_k^c(\varepsilon)\}_{k=1}^{n-1}$ for the full system (see Theorem 3.25) converge to the above positive ones as $\sigma \downarrow 0$. The only one exceptional critical eigenvalue $\lambda_0^c(\varepsilon)$, which belongs to the D^n-symmetric space, does not change its sign when $\sigma \downarrow 0$, because the scalar controller ξ is enough to stabilize the perturbation in this D^n-symmetric space. It is clear from this that spatial variation of the controller v plays a key role in stabilizing multi-layered solutions. We shall study this transition of stability through the analysis of the SLEP matrix $\tilde{M}^{\varepsilon,\sigma}$ as $\sigma \downarrow 0$.

4.1. Instability of Multi-layered Solutions for the Shadow System

We consider the limiting system of (3.1) when $\sigma \downarrow 0$, i.e., the diffusivity of v is extremely large. As $\sigma \downarrow 0$, the second component v approaches a constant function in spatial direction under the assumption that (u, v) remains bounded in C^0-sense for all time which is guaranteed from our assumptions for (f, g) (existence of invariant rectangle). On the other hand, integrating both sides of the second equation of (3.1), and using the Neumann boundary conditions, we have

$$\int_I v_t dx = \int_I g(u, v) dx,$$

which holds independently of σ. Thus we have the following limiting system as $\sigma \downarrow 0$ (see [41] and [30] for more precise derivation);

$$
\begin{aligned}
u_t &= \varepsilon^2 u_{xx} + f(u, \xi) \\
\xi_t &= \int_I g(u, \xi) dx
\end{aligned}
\tag{4.1}
$$

subject to $u_x = 0$ on ∂I, where $\xi(t)$ is a constant function.

We call (4.1) the *shadow system* of (3.1). It is known (see Appendix 1 in [44]) that (4.1) has a unique ε-family of normal n-layered solutions $(u^n(x; \varepsilon, 0), \xi^n(\varepsilon))$, and $D^n(\varepsilon, \sigma) = (u^n(x; \varepsilon, \sigma), v^n(x; \varepsilon, \sigma))$ converges to this solution as $\sigma \downarrow 0$ in $C_\varepsilon^2 \times C^2$-topology. Since the degree of freedom of the controller ξ is equal to one, (4.1) cannot stabilize the normal n-layered solutions $(n \geq 2)$ as observable ones. In terms of the spectral behavior, this can be represented in the next theorem. In this section we only deal with the solutions of (4.1), so we omit the superscript 0 for simplicity like f_v^ε, L^ε, τ_0^* instead of $f_v^{\varepsilon,0}$, $L^{\varepsilon,0}$, $\tau_0^{*,0}$.

Theorem 4.1. *The following linearized eigenvalue problem at* $((u^n(x; \varepsilon, 0), \xi^n(\varepsilon))$

$$\begin{cases} L^\varepsilon w + f_v^\varepsilon \eta = \lambda w \\ \int_I \{g_u^\varepsilon w + g_v^\varepsilon \eta\} = \lambda \eta \end{cases} \qquad (w, \eta) \in (H^2(I) \cap H_N^1(I)) \times \mathbf{C} \qquad (4.2)$$

has exactly n critical eigenvalues $\{\lambda_0^s(\varepsilon), \lambda_1^s(\varepsilon), \cdots, \lambda_{n-1}^s(\varepsilon)\}$ *such that*

$$\lambda_0^s(\varepsilon) < 0 < \lambda_{n-1}^s(\varepsilon) < \cdots < \lambda_1^s(\varepsilon)$$

and satisfy

$$\lambda_0^s(\varepsilon) \simeq \tau_0^* \varepsilon \qquad (\tau_0^* < 0), \qquad (4.3)$$

$$\lambda_i^s(\varepsilon) \leq C \exp(-\gamma/\varepsilon) \qquad (i = 1, \cdots, n-1) \qquad (4.4)$$

when $\varepsilon \downarrow 0$, *where C and γ are positive constants independent of i. All the other noncritical eigenvalues of (4.2) have strictly negative real parts for small ε. Moreover the unstable eigenvalue $\lambda_i^s(\varepsilon)$ coincides with the i-th eigenvalue ζ_i^ε of the Sturm-Liouville operator L^ε for $i = 1, \cdots, n-1$. The associated eigenfunction with $\lambda_0^s(\varepsilon)$ belongs to the D^n-symmetric space X^+, and those for $\{\lambda_i^s(\varepsilon)\}_{i=1}^{n-1}$ are given by $\{(\phi_i^\varepsilon, 0)\}_{i=1}^{n-1}$ and belong to the D^n-breaking space X^-, where ϕ_i^ε is the eigenfunction of L^ε for ζ_i^ε $(i = 1, \cdots, n-1)$. See Lemma 4.3 for the definitions of X^+ and X^-.*

In order to prove this theorem, we need to show the following two lemmas.

Lemma 4.2. *Let* $\{\zeta_i^\varepsilon, \phi_i^\varepsilon\}_{i=0}^\infty$ *be the complete orthnormal set (in L^2-sense) of L^ε. Then ϕ_i^ε has exactly i internal simple zeros and the first n eigenvalues $(\zeta_0^\varepsilon > \zeta_1^\varepsilon > \cdots > \zeta_{n-1}^\varepsilon)$ are positive for $\varepsilon > 0$ and tend to zero exponentially as $\varepsilon \downarrow 0$, i.e.,*

$$0 < \zeta_i^\varepsilon \leq C \exp(-\frac{\gamma}{\varepsilon}) \qquad i = 0, 1, \cdots, n-1, \qquad (4.5)$$

where C and γ are positive constants independent of i.

Proof. The proof of Lemma 3.15 is also valid for this case. Since $v = \xi$ is a constant function, the first term of (3.44) does not appear, which leads us to the estimate (4.5). The details are left to the reader.

Next we introduce the orthogonal decomposition of $L^2(I)$ which takes into account the symmetry of normal n-layered solutions due to the folding up principle (see Corollary 3.9). A function $u(x) \in L^2(I)$ is called D^n-*symmetric* if it is generated by $u(x)|_{[0,1/n]}$

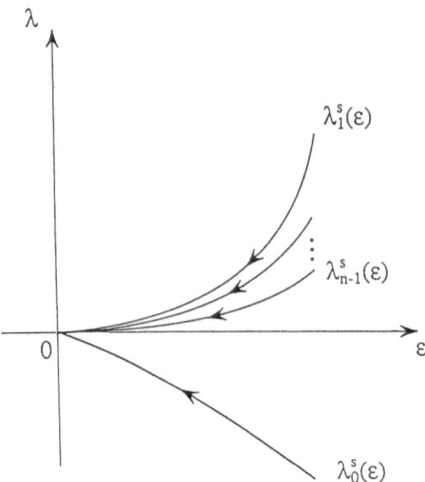

Figure 4.1.

(restriction to $[0, 1/n]$) by flipping it $(n - 1)$-times in an even way. Apparently each component of the normal n-layered solution is D^n-symmetric. $L^2(I)$ can be decomposed into D^n-symmetric part and its orthogonal complement in the following way.

Lemma 4.3. *The space $L^2(I)$ has the following orthogonal decomposition*

$$L^2(I) = X^+ \oplus X^-,$$

where

$$X^+ \equiv closure\ of\ span\{\phi^\varepsilon_{nk}\}^\infty_{k=0} \qquad in\ L^2(I)$$

$$X^- \equiv closure\ of\ span\{\phi^\varepsilon_i\}_{i\neq nk} \qquad in\ L^2(I).$$

X^+ consists of the D^n-symmetric functions, in paticular, normal n-layered solutions and constant function belong to this space. We call X^+ (resp. X^-) the D^n-symmetric (resp. D^n-breaking)-space. Finally we denote the orthogonal projection onto X^+ by P^+.

Proof. Since the potential term f^ε_u of L^ε is D^n-symmetric and from the nodal property of eigenfunctions, $\{\phi^\varepsilon_{nk}\}^\infty_{k=0}$ is generated via folding from a complete orthogonal base of the restricted Sturm-Liouville problem L^ε to $(0, 1/n)$ with Neumann boundary conditions at $x = 0$ and $1/n$. This immediately leads us to the above lemma. $\square$

Remark 4.4. *Let $\tilde{L}^2(I)$ be an extended space of $L^2(I)$ defined by*

$$\tilde{L}^2(I) \equiv \{each\ element\ of\ L^2(I)\ is\ extended\ to\ the\ double\ interval\ (0, 2)$$

in an even way and identify both end points}.

Similar extension can be applied to X^+ and X^-, and we have an orthogonal decomposition

$$\bar{L}^2(I) = \tilde{X}^+ \oplus \tilde{X}^-.$$

Then the element of $\tilde{X}^+$ is characterized as a function which is invariant under D^n-action, i.e., reflection and $2\pi/n$-rotation. Note that X^+ is invariant under usual algebraic operations such as addition, product and so on. For more systematic group theoretical approach to our problems, see [24] and [25].

Proof of Theorem 4.1. The eigenvalue problem (4.2) can be decomposed as follows by using the projection P^+ onto X^+:

$$\begin{cases} L^\varepsilon w^+ + f_v^\varepsilon \eta = \lambda w^+ \\ \\ \int (g_u^\varepsilon w^+ + g_v^\varepsilon \eta) dx = \lambda \eta \end{cases} \tag{4.6}$$

$$\begin{cases} L^\varepsilon w^- = \lambda w^- \\ \\ \int g_u^\varepsilon w^- \, dx = 0, \end{cases} \tag{4.7}$$

where $w = w^+ + w^-$, $w^+ \equiv P^+ w \in X^+$, and $w^- = (I - P^+)w \in X^-$. Here we use the fact that L^ε commutes with P^+, and that f_v^ε, g_u^ε, g_v^ε, and η belong to X^+. The system (4.7) is equivalent to

$$L^\varepsilon w^- = \lambda w^-, \tag{4.8}$$

since w^- ($\in X^-$) is orthogonal to g_u^ε ($\in X^+$). Since each component of the decomposition

$$L^2(I) \times \mathbf{R} = (X^+ \times \mathbf{R}) \oplus (X^- \times \{0\})$$

is invariant under the linearized operation (i.e., the left-hand side of (4.2)). It suffices to solve problems (4.6) and (4.8) separately.

First it is clear from (4.8), Lemmas 4.2 and 4.3 that the set of eigenvalues $\{\zeta_i^\varepsilon\}_{i=1}^{n-1}$ of L^ε are the required positive critical eigenvalues $\{\lambda_i^s(\varepsilon)\}_{i=1}^{n-1}$ in Theorem 4.1 and all the other spectra of (4.8) are noncritical and strictly negative for small ε. The remaining negative critical eigenvalue $\lambda_0^s(\varepsilon)$ can be obtained by solving (4.6) in the following way. Using Lemma 4.2, the first equation of (4.6) can be solved with respect to w^+:

$$\begin{aligned} w^+ &= (L^\varepsilon - \lambda)^{-1}(-f_v^\varepsilon \eta) \\ \\ &= \frac{\langle -f_v^\varepsilon \eta, \phi_0^\varepsilon \rangle}{\zeta_0^\varepsilon - \lambda} \phi_0^\varepsilon + (L^\varepsilon - \lambda)^\dagger (-f_v^\varepsilon \eta), \end{aligned} \tag{4.9}$$

where $(L^\varepsilon - \lambda)^\dagger \equiv \displaystyle\sum_{k=1}^\infty \frac{\langle -f_v^\varepsilon \eta, \phi_{kn}^\varepsilon \rangle}{\zeta_{kn}^\varepsilon - \lambda} \phi_{kn}^\varepsilon$, the reduced resolvent defined on X^+, which satisfies all the properties in Section 3, especially Lemma 3.19. Also note that ζ_0^ε does not belong to the spectrum of (4.6) for small ε, which can be proved in an analogous way of Lemma 3.17. Substituting (4.9) into the second equation of (4.6), we have after dividing it by η

$$\int_I \left\{ \frac{(-f_v^\varepsilon, \phi_0^\varepsilon)}{\zeta_0^\varepsilon - \lambda} g_u^\varepsilon \phi_0^\varepsilon + g_u^\varepsilon (L^\varepsilon - \lambda)^\dagger (-f_v^\varepsilon) + g_v^\varepsilon \right\} dx = \lambda. \qquad (4.10)$$

When $\eta = 0$, it follows from (4.9) that $w \equiv 0$, hence it suffices to consider (4.10).

Suppose λ is a noncritical eigenvalue. It follows from Corollary 3.14 and the fact that the denominator is bounded away from zero that the first term on the left-hand side of (4.10) tends to zero. In view of Lemma 3.19, we see that the second term of (4.10) approaches $\frac{-g_u^* f_v^*}{f_u^* - \lambda}$ as $\varepsilon \downarrow 0$. Thus (4.10) becomes in the limit of $\varepsilon \downarrow 0$

$$\int_I \left\{ \frac{\det^* - \lambda g_v^*}{f_u^* - \lambda} \right\} dx = \lambda, \qquad (4.11)$$

where $\det^* \equiv f_u^* g_v^* - g_u^* f_v^*$. Recalling the assumptions (A.3) $\sim$ (A.5), one can show after some computation that any solution λ of (4.11) must have strict negative real part, which is also true for small positive ε by continuity arguments. Thus noncritical eigenvalues are not dangerous to stability.

We next consider the case where λ is a critical eigenvalue. Since the same asymptotic characterization of Lemma 3.22 and 3.23 also holds for the shadow system, we rewrite it in the following form

$$\int_I \left\{ \frac{(-f_v^\varepsilon, \phi_0^\varepsilon / \sqrt{\varepsilon})}{\hat{\zeta}_0^\varepsilon - \tau} g_u^\varepsilon \frac{\phi_0^\varepsilon}{\sqrt{\varepsilon}} + g_u^\varepsilon (L^\varepsilon - \varepsilon\tau)^\dagger (-f_v^\varepsilon) + g_v^\varepsilon \right\} dx = \varepsilon\tau, \qquad (4.12)$$

where $\tau \equiv \lambda/\varepsilon$ and $\hat{\zeta}_0^\varepsilon \equiv \zeta_0^\varepsilon / \varepsilon$. Note that $\hat{\zeta}_0^\varepsilon$ tends to zero exponentially (see Lemma 4.1). Hence, when $\varepsilon \downarrow 0$, (4.12) becomes

$$-\frac{c_1^* c_2^*}{\tau^*} + \int_I \frac{\det^*}{f_u^*} dx = 0,$$

which leads us to the expression

$$\tau_0^* \equiv \lim_{\varepsilon \downarrow 0} \tau(\varepsilon) = \frac{c_1^* c_2^*}{\displaystyle\int_I \left(\frac{\det^*}{f_u^*} \right) dx}. \qquad (4.13)$$

Recalling (A.3) and (A.4b), we see that $\tau_0^* < 0$. Multiplying $\hat{\zeta}_0^\varepsilon - \tau$ on both sides of (4.12) and applying the implicit function theorem to it, we easily see that (4.12) has a unique continuous solution $\tau = \tau(\varepsilon)$ up to $\varepsilon = 0$ with $\tau(0) = \tau_0^*$ given by (4.13). This completes the proof of Theorem 4.1. $\square$

4.2. Recovery of Stability (From the shadow system to the full system)

We shall fill the gap of the stability results between the shadow system (Theorem 4.1) and the full system (Theorem 3.27) by studying the eigenvalue problem (3.66) in the limit of $\sigma \downarrow 0$. The next lemma is a key observation for this purpose.

Lemma 4.5. *When* $\sigma \downarrow 0$, *the operator* $K^{\varepsilon,\sigma,\lambda}$ *defined in Lemma 3.21 converges to* $K^{\varepsilon,0,\lambda}$ *in operator norm sense, where* $K^{\varepsilon,0,\lambda}$ *maps* $h \in H^{-1}(I)$ *to the constant function* $c(h)$ *defined by*

$$c(h) \equiv -\langle h, 1 \rangle \Big/ \langle g_u^{\varepsilon}(L^{\varepsilon} - \lambda)^{\dagger}(-f_v^{\varepsilon}) + g_v^{\varepsilon} - \lambda, 1 \rangle . \tag{4.14}$$

For $\lambda = 0$, *(4.14) tends to the following as* $\varepsilon \downarrow 0$

$$c^*(h) = -\frac{\langle h, 1 \rangle}{\langle \det^* /f_u^*, 1 \rangle}. \tag{4.15}$$

Idea of proof. Although we refer the proof to that of Lemma 3.1 of [44], we give here an intuitive idea. Roughly speaking, $K^{\varepsilon,\sigma,\lambda}$ can be expressed by

$$K^{\varepsilon,\sigma,\lambda} = \left\{ -\frac{1}{\sigma}\frac{d^2}{dx^2} - g_u^{\varepsilon,\sigma}(L^{\varepsilon,\sigma} - \lambda)^{\dagger}(-f_v^{\varepsilon,\sigma}\cdot) - g_v^{\varepsilon,\sigma} + \lambda \right\}^{-1}.$$

We decompose h as

$$h = \hat{h} + \int_I h\,dx,$$

where the average of $\hat{h}$ is equal to zero, i.e., $\int_I \hat{h}\,dx = 0$. According to this orthogonal decomposition, the operator $K^{\varepsilon,\sigma,\lambda}$ is splitted into two parts; one acts on $\hat{h}$-space and the other on average-space. As $\sigma \downarrow 0$, $K^{\varepsilon,\sigma,\lambda}\big|_{\hat{h}-space}$ tends to zero in operator norm because of $\left\{ -\frac{1}{\sigma}\frac{d^2}{dx^2} \right\}^{-1}$. Hence only the average part remains nontrivial which is determined by

$$-c(h)\int_I \left\{ g_u^{\varepsilon,\sigma}(L^{\varepsilon,\sigma} - \lambda)^{\dagger}(-f_v^{\varepsilon,\sigma}) + g_v^{\varepsilon,\sigma} - \lambda \right\} dx = \int_I h\,dx.$$

This leads us to (4.14). $\square$

Now we can prove the following.

Theorem 4.6. *When* $\sigma \downarrow 0$, *the eigenvalues of (3.66)*

$$\bar{M}^{\varepsilon,\sigma}\mathbf{a} = \begin{pmatrix} (\zeta_0^{\varepsilon,\sigma}/\varepsilon - \lambda/\varepsilon)\alpha_0 \\ \cdot \\ \cdot \\ \cdot \\ (\zeta_{n-1}^{\varepsilon,\sigma}/\varepsilon - \lambda/\varepsilon)\alpha_{n-1} \end{pmatrix} \tag{3.66}$$

tend to the critical eigenvalues $\{\lambda_0^s(\varepsilon), \lambda_1^s(\varepsilon), \cdots, \lambda_{n-1}^s(\varepsilon)\}$ *for the shadow system in Theorem 4.1.*

Proof. The (i, j)-component of $\bar{M}^{\varepsilon,\sigma}$ is equal to $\langle -f_v^{\varepsilon,\sigma}\phi_i^{\varepsilon,\sigma}/\sqrt{\varepsilon}, K^{\varepsilon,\sigma,\lambda}(g_u^{\varepsilon,\sigma}\phi_j^{\varepsilon,\sigma}/\sqrt{\varepsilon})\rangle$ (see (3.66)). In view of Lemma 4.3, we see that both $-f_v^{\varepsilon,\sigma}\phi_i^{\varepsilon,\sigma}/\sqrt{\varepsilon}$ and $g_u^{\varepsilon,\sigma}\phi_j^{\varepsilon,\sigma}/\sqrt{\varepsilon}$ are orthogonal to constant function for $i, j \neq 0$, i.e., their averages are equal to zero, while

$\int_I g_u^{\varepsilon,\sigma}(\phi_0^{\varepsilon,\sigma}/\sqrt{\varepsilon})dx$ converges to a nonzero constant as $\sigma \downarrow 0$. Hence all components except $(1, 1)$ of $\tilde{M}^{\varepsilon,\sigma}$ approach zero from Lemma 4.5 when $\sigma \downarrow 0$. More precisely, we have

$$\tilde{M}^{\varepsilon,0} \equiv \lim_{\sigma\downarrow0} \tilde{M}^{\varepsilon,\sigma} = \begin{pmatrix} m(\varepsilon, \lambda) & \cdot & \cdot & \cdot & 0 \\ \cdot & & & & \\ \cdot & & O & & \\ \cdot & & & & \\ 0 & & & & \end{pmatrix} \tag{4.16a}$$

where

$$m(\varepsilon, \lambda) \equiv \left\langle -f_v^\varepsilon \phi_0^\varepsilon/\sqrt{\varepsilon}, - \int_I g_u^\varepsilon(\phi_0^\varepsilon/\sqrt{\varepsilon})dx \Big/ \int_I \{g_u^\varepsilon(L^\varepsilon - \lambda)^\dagger(-f_v^\varepsilon) + g_v^\varepsilon - \lambda\}dx \right\rangle \tag{4.16b}$$

and

$$m(0, 0) = -c_1^* c_2^* \Big/ \int_I (\det{}^*/f_u^*)dx . \tag{4.16c}$$

In view of the eigenvalue problems (3.66), (3.67), and the proof of Theorem 4.1 (see (4.12) and (4.13)), the above asymptotic characterization (4.16) leads us to Theorem 4.6. □

Remark 4.7. *In Theorem 4.6, we used (3.66) which may depend on the choice of the subsequence, however the proof of it does not depend on such a choice. In fact, the principal eigenfunction $\phi_0^{\varepsilon,\sigma}/\sqrt{\varepsilon}$ is itself a convergent sequence and the orthogonal property used in the proof of Theorem 4.6 holds for any subsequence.*

5. Concluding Remarks

Up until now we have considered the reaction-diffusion systems (1.1) and applied the SLEP method to the layered solutions to study the stability properties. The basic idea of this method, however, has a wide range of applicability to various types of problems, which essentially comes from the universal structure of internal layers of this class. In what follows we shall discuss briefly about several topics to which the SLEP method is useful.

(i) *Systems with different scales of relaxation parameters*

If the relaxation parameter δ of (1.1a) is taken to be $\varepsilon\tau$ ($\tau = O(1)$), then the dynamics of (1.1) and its singular limit system drastically change. In fact, it is easily seen from (1.11) that the propagation speed of the internal layer is of $O(1)$ and no longer slow compared with the outer dynamics. A similar asymptotic analysis as in Section 1 yields the following singular limit dynamics:

$$(\varphi_i)_t = \frac{(-1)^{i-1}}{\tau} c(V(\varphi_i(t))) \tag{5.1a}$$

$$V_t = DV_{xx} + G_\Phi(V). \tag{5.1b}$$

The dynamics of layers ((5.1a)) evolves simultaneously with that of the outer part ((5.1b)). It is plausible that, when τ becomes small, the outer dynamics cannot catch up with the speed of layers, and hence is not able to settle them down to a steady state. This instability really occurs in the form of Hopf bifurcation for the original system (1.1) with $\delta = \varepsilon\tau$ as well as for (5.1), and we have *layer oscillations* (*breathers*) (see [46] and [42] for details). The well-posedness and asymptotic behaviors of (5.1) has been proven by [30] for mono-layer case. For multi-layer case, the dynamics becomes more rich and complicated such as synchronization, annihilation, and coalescence, the study of which is under progress (see [31] for the case of two breathers). Finally, suppose we consider the system (1.1) with $\delta = \varepsilon\tau$ on the entire line **R**, instability also occurs for travelling front solutions as τ becomes small, however the structure of bifurcation is different from the finite interval case due to the translation invariance. See [47] and [35] for details.

(ii) *Neumann layered solutions and separators*

The Main Theorem in Section 1 shows the coexistence of arbitrary many stable steady states in the limit of $\varepsilon \downarrow 0$, however, in order for the system (1.1) to be self-consistent, there must exist other unstable steady states or invariant sets which play the role of *separators* among stable ones. In fact it is possible to construct such unstable layer solutions in a rather systematic way (see Fujii and Hosono [23], and Nishiura and Tsujikawa [50]) although it may not exhaust all types of unstable solutions. Typical profiles are illustrated in Figure 5.1. These solutions consist of normal n-layered solutions plus *Neumann layers* (i.e., boundary layers satisfying the Neumann boundary condition), and play the role of separators among normal layered solutions. For example, it is not difficult to imagine that the solution in Figure 5.1 (a) is the separator between normal 1-layer and 2-layer solutions (see Figure 5.2). More precisely, the dimension of the unstable manifold of the Neumann layered solution is equal to one, and it is connected to 1-layer and 2-layer solutions. However, in general, rigorous justification of the existence of these connecting orbits remains an unexplored field compared with the scalar case (see Brunovský and Fiedler [5] and references therein). This is partly because of the lack of powerful tools such as Lyapunov function and the lap number for the system (1.1).

(iii) *Heteroclinic and Homoclinic bifurcations*

Global bifurcation such as heteroclinic or homoclinic bifurcation is one of the most important issues in dynamical system theory, and one can find many applications to various fields. For instance, a creation of homoclinic loop from two heteroclinic orbits is quite interesting from a PDE view point, since it means that travelling pulse solutions are born from a pair of front and back solutions (see, for example, Rinzel and Terman [54]). However, when one tries to apply such tools to practical problems, one obviously has to check several transversal and generic conditions along *large amplitude* orbits (see, for example, Chow, Deng, and Terman [13], Deng [14], and Kokubu [34]). This is usually not an easy task without a good control of parametric dependency of generating orbits from which new kind of solutions emanate when parameters vary. The SLEP method is suitable for this purpose when such orbits can be constructed by singular perturbation, since the usual transversal conditions are related to the spectral behavior of the linearized problem at the generating orbits. In fact all the hypotheses imposed on heteroclinic and homoclinic bifurcation theorem are rigorously verified by Kokubu, Nishiura, and

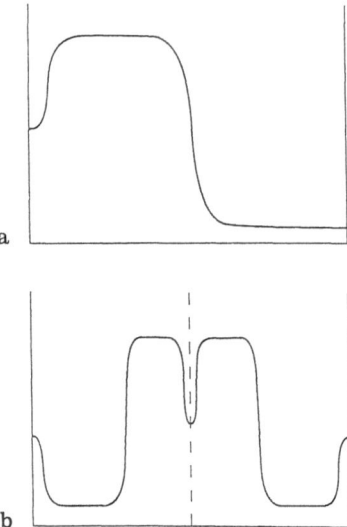

a

b

Figure 5.1.

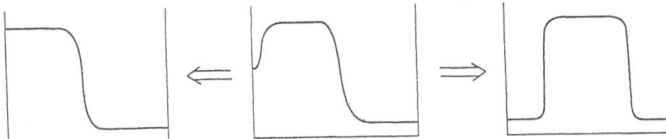

Figure 5.2.

Oka [35] for a system of bistable reaction diffusion equations with the aid of the SLEP method. A relation between the stability of front (or back) solutions and the intersecting manner of the stable and unstable manifolds is also given in [35].

(iv) *Higher dimensional cases*

The study of morphologies of interfaces in higher space dimensions such as dendrites in solidification problem (Langer [36]) and spirals in chemical reaction (Fife [19], Keener and Tyson [33]) is a central problem in pattern formation theory. Such phenomena can be modelled by reaction diffusion systems (see, for instance, Caginalp [6] and Fife [20]) containing a small parameter ε which represents the width of interface, and interfacial patterns can be constructed as singularly perturbed solutions to the model systems. Physically speaking, this singular perturbation could be explained as the introduction of surface tension effect (see Pelcé [53]). Stability analysis and bifurcation in these interfacial problems are quite important, since they are directly related to *pattern selection* problem. Unfortunately there are very few works on this issue, partly because it is, in general, extremely difficult to show the existence of the singularly perturbed solutions in a constructive way in higher dimensional space (for a single equation, see Fife and Greenlee [21]). For special domains like channels or spherical shapes, it is possible to answer, to some extent, for both existence and stability for reaction diffusion systems (see

Ohta, Mimura, and Kobayashi [52], Ohta and Mimura [51], and Taniguchi and Nishiura [57]). To proceed further, it seems reasonable to assume, as a working hypothesis, the existence of an ε-family of singularly perturbed solutions which has a smooth interface as $\varepsilon \downarrow 0$. Then the basic question is that how one can characterize the stability or instability of an interface by the *geometry* of it and *outer* solutions. And is it possible to derive the singular limit eigenvalue problem contracted on the interface?

Let us consider the reaction diffusion system in $\Omega \subset \mathbf{R}^n$:

$$
\begin{cases}
u_t = \varepsilon^2 \Delta u + f(u, v) & \\
& \quad \text{in } \Omega \subset \mathbf{R}^n \\
v_t = D\Delta v + g(u, v) & \\
\dfrac{\partial u}{\partial n} = 0 = \dfrac{\partial v}{\partial n} & \quad \text{on } \partial\Omega,
\end{cases}
\tag{5.2}
$$

where Ω is a smooth bounded domain and $\partial/\partial n$ denotes the normal derivative, and assume that it has an ε -family of singularly perturbed steady state solutions with a smooth closed hypersurface Γ_0 as $\varepsilon \downarrow 0$. It turns out that the singular limit eigenvalue problem on Γ_0 is given by

$$
\varepsilon\{\Delta_s + H \cdot\}\gamma + \frac{dc}{dv}(V^*\big|_{\Gamma_0})\{(\nabla V^* \cdot n)\gamma + [G]\langle K^*(\delta_{\Gamma_0} \otimes \gamma), \delta_{\Gamma_0}\rangle\} = \tau\gamma, \tag{5.3}
$$

where Δ_s represents the Laplace operator on Γ_0, $H \cdot$ is a bounded operator on $L^2(\Gamma_0)$ which depends on the geometry of Γ_0 but not on ε, $c(V)$ the velocity function as in (1.9), V^* the outer solution for v, $[G]$ the jump of value of g at Γ_0, δ_{Γ_0} the surface distribution of *Dirac*'s δ on Γ_0, K^* a compact operator similar to $K^{*,\sigma,0}$ (see (2.27) and Lemma 3.21), $\tau(\equiv \lim_{\varepsilon\downarrow 0} \lambda/\varepsilon)$ the scaled eigenvalue as in Lemma 3.23, and γ is an eigenfunction in $L^2(\Gamma_0)$ (see Nishiura [43] for a formal derivation of (5.3)). We call (5.3) *the SLEP equation on* Γ_0 for (5.2). Apparently (5.3) is itself a singular perturbation problem because of the first term on the left-hand side. The term $\varepsilon\Delta_s\gamma$ *cannot be neglected* since high frequency modes are stabilized by this term. It should be noted that the nonlocal term $[G]\langle K^*(\delta_{\Gamma_0} \otimes \gamma), \delta_{\Gamma_0}\rangle$ is responsible for the stabilization of low frequency modes. On the other hand, the second term $(\nabla V^* \cdot n)\gamma$ on the left-hand side is the principal part of destabilizing effect. Therefore only a finite band of modes could be destabilized in this system. In fact it can be proved that *any* stationary pattern of (5.2) with *smooth* limiting interface becomes *unstable* for small ϵ. Then what stable patterns look like and how they behave when ϵ tends to zero? One possibility is that the characteristic domain size of stable stationary patterns tends to zero as $\epsilon \downarrow 0$. This implies that stable patterns become finer and finer as $\epsilon \downarrow 0$. WE shall discuss more on these issues in Nishiura and Suzuki [49].

Appendix A

Proof of Lemma 2.7.

First we prove the equality

$$-\frac{dc}{dV}(v^*) = \frac{1}{\left\|\dfrac{d}{dy}\tilde{u}^*\right\|_{L^2}^2}\frac{dJ}{dv}(v^*). \tag{1}$$

It follows from (1.8) and (1.9) that

$$u_{yy} + c(V)u_y + f(u, V) = 0 \tag{2a}$$

$$u(\pm\infty) = h_{\pm}(V). \tag{2b}$$

Differentiating (2a) by V, we have

$$(u')_{yy} + c(V)(u')_y + f_u(u, V) = -\frac{dc}{dV}(V)u_y - f_v(u, V)$$

where $u' = \dfrac{du}{dV}$. Let $V = v^*$, then

$$(u')_{yy} + f_u(u, v^*)u' = -\frac{dc}{dV}(v^*)u_y - f_v(u, v^*). \tag{3}$$

Here we used $c(v^*) = 0$. On the other hand, recalling Remark 3.7, $\dfrac{d}{dy}\tilde{u}^*$ satisfies

$$W_{yy} + f_u(W, v^*)W = 0$$

with $\|W\|_{L^2} < \infty$, i.e., $\dfrac{d}{dy}\tilde{u}^*$ belong to the kernel of the operator $\dfrac{d^2}{dy^2} + f_u(W, v^*)$ in $L^2(\mathbf{R})$. Applying the solvability condition to (3), we have

$$-\frac{dc}{dV}(v^*)\left\|\frac{d}{dy}\tilde{u}^*\right\|^2 = \int_{\infty}^{\infty} f_v(\tilde{u}^*, v^*)\frac{d}{dy}\tilde{u}^* dy$$

$$= \frac{d}{dv}\int_{h_-(v^*)}^{h_+(v^*)} f(u, v^*)du,$$

which shows (1).

Recalling the definitions of constants γ^*, c^*, c_1^* and c_2^* (see Remark 3.7 and Lemma 3.22), (1) implies (ii) of Lemma 2.7. As for (i), using the fact that V_2^* satisfies

$$\frac{1}{\sigma}(V_2^*)_{xx} + g(U_2^*, V_2^*) = 0$$

$$(V_2^*)_x = 0 \qquad\qquad \text{at} \quad x = 0, 1,$$

it is easily seen that

$$\frac{dV_2^*}{dx}(\varphi_1^*) = -\sigma \int_0^{\varphi_1^*} g(U_2^*, V_2^*)dx. \tag{4}$$

Combining (1), (4), with Lemma 3.15, we obtain (i) of Lemma 2.7. $\square$

Appendix B

Proof of Lemma 3.17.

We shall prove a slightly more general statement: There are no eigenvalues of $\mathcal{L}^{\varepsilon,\sigma}$ which have the same asymptotic behaviour as (3.39) of Lemma 3.15. Let us prove this by contradiction. Suppose that there exists an eigenvalue $\lambda = \lambda(\varepsilon)$ of $\mathcal{L}^{\varepsilon,\sigma}$ such that

$$\lim_{\varepsilon \downarrow 0} \frac{\lambda(\varepsilon)}{\varepsilon} = \xi^{*,\sigma} \tag{1}$$

with the associated eigenfunction denoted by $(w(\varepsilon), z(\varepsilon))$. Hereafter we simply write $\lambda(\varepsilon)$, $w(\varepsilon)$, $z(\varepsilon)$ as λ, w, z. Solving the first equation of (3.12a) with respect to w, we have

$$w = \sum_{j=0}^{n-1} k_j(\varepsilon; z)\phi_j^{\varepsilon,\sigma} + (L^{\varepsilon,\sigma} - \lambda)^\dagger(-f_v^{\varepsilon,\sigma}z), \tag{2}$$

where

$$k_j(\varepsilon; z) \equiv \begin{cases} \dfrac{\langle -f_v^{\varepsilon,\sigma}z, \phi_j^{\varepsilon,\sigma}\rangle}{\zeta_j^{\varepsilon,\sigma} - \lambda} & \text{if } \lambda \neq \zeta_j^{\varepsilon,\sigma} \\ \\ c_j & \text{arbitrary constant if } \lambda = \zeta_j^{\varepsilon,\sigma}. \end{cases} \tag{3}$$

Note that if $\lambda = \zeta_j^{\varepsilon,\sigma}$, the solvability condition

$$\langle -f_v^{\varepsilon,\sigma}z, \phi_j^{\varepsilon,\sigma}\rangle = 0 \tag{4}$$

must be satisfied. Also, if necessary, by choosing an appropriate subsequence of $\lambda(\varepsilon)$ (we use the same notation for this), we can assume without loss of generality that either one of the cases of (3) occur when $\varepsilon \downarrow 0$. In what follows we treat only the first case since the second case can be dealt with in a similar way by using (4). Substituting (2) into the second equation of (3.12a) and applying $K^{\varepsilon,\sigma,\lambda}$ of Lemma 3.21 to it, we obtain

$$z = K^{\varepsilon,\sigma,\lambda}\left(\sum_{j=0}^{n-1} \sqrt{\varepsilon}k_j(\varepsilon; z)g_u^{\varepsilon,\sigma}\frac{\phi_j^{\varepsilon,\sigma}}{\sqrt{\varepsilon}}\right). \tag{5}$$

It is clear that not all k_j's are zero, otherwise $(w, z) \equiv 0$. Without loss of generality, we can normalize z as

$$\| z \|_{H_N^1(I)} = 1. \tag{6}$$

Recalling Lemma 3.21 and 3.22, we see as a necessary condition of (6) that $\sqrt{\varepsilon}k_j$ are uniformly bounded and not all of them go to zero as $\varepsilon \downarrow 0$. In view of (3), we see that

$$\sqrt{\varepsilon}\langle -f_v^{\varepsilon,\sigma}z, \phi_j^{\varepsilon,\sigma}\rangle = o(\varepsilon) \tag{7}$$

holds because $|\zeta_j^{\varepsilon,\sigma} - \lambda| = o(\varepsilon)$ as $\varepsilon \downarrow 0$ from our assumption (1). Dividing (7) by ε, we have

$$\lim_{\varepsilon \downarrow 0}\langle -f_v^{\varepsilon,\sigma}z, \frac{\phi_j^{\varepsilon,\sigma}}{\sqrt{\varepsilon}}\rangle = 0 \qquad\qquad \text{for all } j. \tag{8}$$

Choosing an appropriate subsequence of $\sqrt{\varepsilon}k_j$ (with keeping the same notation), we obtain

$$\sqrt{\varepsilon}k_j(\varepsilon;z) \longrightarrow \hat{k}_j^* \qquad\qquad \text{for all } j.$$

and the vector $\hat{\mathbf{k}}^* \equiv (\hat{k}_0^*, \cdots, \hat{k}_{n-1}^*)$ is not equal to zero. Therefore we have in the limit of $\varepsilon \downarrow 0$,

$$z^* \equiv \lim_{\varepsilon \downarrow 0} z(\varepsilon) = K^{*,\sigma,0}\left(c_2^* \sum_{j=0}^{n-1} \hat{k}_j^*\Delta_j\right). \tag{9}$$

On the other hand, it follows from (8) that

$$\langle z^*, \Delta_j\rangle = 0 \qquad\qquad \text{for all } j,$$

which implies

$$\langle z^*, \sum_{j=0}^{n-1} \hat{k}_j^*\Delta_j\rangle = 0. \tag{10}$$

Substituting the expression (9) into (10), we have

$$\langle K^{*,\sigma,0}\hat{w}^*, \hat{w}^*\rangle = 0, \tag{11}$$

where $\hat{w}^* = \sum_{j=0}^{n-1} \hat{k}_j^*\Delta_j$. Using the positive definiteness of $K^{*,\sigma,0}$ (see Lemma 3.21), $\hat{w}^* = 0$ from (11), and hence $z^* = 0$, which is a contradiction.

Appendix C

First we list up four sublemmas necessary to prove Lemmas 3.28 and 3.32. Recalling (3.69a) and (3.71), we can write $G_N = -\dfrac{\sigma}{W(h, k)}G$, where G is a real symmetric matrix defined by $G = \{h_i k_j\}_{i,j=1}^n$ with $h_i = h(x_i^*(\sigma))$ and $k_j = k(x_j^*(\sigma))$. It suffices to consider the inverse of G for the proof.

Sublemma 1. *Let G be the matrix $\{h_i k_j\}_{i,j=1}^n$ defined as above and let ΔG_{ij} be the (i,j)-cofactor of G. Then we have*

$$\det G = (-1)^n A_{12} \cdot A_{23} \cdots A_{n-1,n} \cdot h_1 k_n \tag{a}$$

$$\Delta G_{ii} = (-1)^{n-2} A_{12} \cdots A_{i-1,i+1} \cdot A_{i+1,i+2} \cdots A_{n-1,n} \cdot h_1 k_n, \qquad i \neq 1, n \qquad (b)$$

$$\Delta G_{11} = (-1)^{n-2} A_{23} \cdots A_{n-1,n} \cdot h_2 k_n \qquad (c)$$

$$\Delta G_{nn} = (-1)^{n-2} A_{12} \cdots A_{n-2,n-1} \cdot h_1 k_{n-1} \qquad (d)$$

$$\Delta G_{i+1,i} = (-1)^{n-1} A_{12} \cdots \check{A}_{i,i+1} \cdots A_{n-1,n} \cdot h_1 k_n, \qquad 1 \leq i \leq n-1 \qquad (e)$$

$$\Delta G_{ij} = 0 \qquad\qquad (i,j) \notin \text{tri-diagonal,} \qquad (f)$$

where

$$A_{ij} \equiv \begin{vmatrix} h_i & k_i \\ h_j & k_j \end{vmatrix}$$

and $\check{}$ *means that the right-hand side of (e) lacks its term.*

Sublemma 2.

$$A_{ij} = \begin{vmatrix} h_i & k_i \\ h_j & k_j \end{vmatrix} < 0, \qquad\qquad (i < j) \qquad (a)$$

$$A_{ij} = A_{i+2,j+2} \qquad (b)$$

$$A_{13} = A_{24} \qquad (c)$$

Sublemma 3.

$$\frac{W(h,k)}{A_{12}} = \frac{W(h^d, k^d)}{A_{12}^d} \qquad (a)$$

$$\frac{W(h,k)}{A_{23}} = \frac{W(h^d, k^d)}{A_{23}^d} \qquad (b)$$

$$\frac{W(h,k)}{A_{13}} = \frac{W(h^d, k^d)}{A_{13}^d} \qquad (c)$$

Sublemma 4.

$$\frac{h_2}{h_1} < \frac{h_2^d}{h_1^d} \qquad (a)$$

$$\frac{k_{n-1}}{k_n} < \frac{k_{n-1}^d}{k_n^d} \qquad (b)$$

Proof of Lemma 3.28.
The existence of G_N^{-1} is clear from Sublemma 1(a) and Sublemma 2(a). Since G_N is real symmetric, so is G_N^{-1}. Tri-diagonality is a direct consequence of Sublemma 1(f). Using Sublemma 1, it is clear that G_N^{-1} takes the form (3.86) or (3.87). Properties (a)~(c) are the direct consequences of Sublemma 2.

Proof of Lemma 3.32.
Sublemmas 3 and 4 imply Lemma 3.32.

In what follows we shall prove Sublemmas 1 - 4.

Proof of Sublemma 1.
We prove only (a) and (f). The remaining part can be shown in a similar way. Recall that G takes the form

$$G \equiv \begin{pmatrix} h_1 k_1 & h_1 k_2 & h_1 k_3 & \cdots & h_1 k_n \\ h_1 k_2 & h_2 k_2 & h_2 k_3 & \cdots & h_2 k_n \\ h_1 k_3 & h_2 k_3 & h_3 k_3 & \cdots & h_3 k_n \\ \vdots & \vdots & \vdots & \ddots & \\ h_1 k_n & & & & h_n k_n \end{pmatrix} \tag{1}$$

Note that although the original (i, j)-component of G for $i > j$ is given by $h_j k_i$, we rewrite it as above by using the symmetry of G. Hence, at the i-th row, the subscript of h increases up to the diagonal part and keeps the number i after that like $(1, 2, \cdots, i, i, \cdots, i)$. We shall make G into a triangular matrix to compute its determinant. Dividing each i-th row by h_i, and substracting from the i-th column the $(i+1)$-th column multiplied by h_i/h_{i+1}, we obtain an upper triangular matrix whose product of diagonal elements gives

$$\det G = \left(k_1 - \frac{h_1 k_2}{h_2}\right)\left(k_2 - \frac{h_2 k_3}{h_3}\right)\cdots\left(k_{n-1} - \frac{h_{n-1} k_n}{h_n}\right) k_n \times (h_1 \cdots h_n)$$

$$= (-1)^{n-1} A_{12} A_{23} \cdots A_{n-1,n} \cdot h_1 \cdot k_n,$$

which shows (a).
As for (f), it suffices to consider the case $i < j$. Let G_{ij} denote the submatrix of G which lacks i-th row and j-th column. We decompose G_{ij} into four block matrices as

$$G_{ij} = \begin{pmatrix} B_1 & B_3 \\ B_2 & B_4 \end{pmatrix},$$

where B_1 is a square matrix of size $j - 1$. Dividing i-th row by h_i $(1 \le i \le j - 1)$, we easily see that B_3 can be reduced to zero matrix by fundamental transformations. Moreover the last two row vectors of B_1 are linearly dependent, since they are of the form

$$j - 2 : \frac{k_{j-1}}{h_{j-1}}(h_1, h_2, \cdots, h_{j-1})$$

$$j - 1 : \frac{k_j}{h_j}(h_1, h_2, \cdots, h_{j-1}).$$

Combining these two results, we can conclude (f). □

In order to prove Sublemmas 2-4, we need to prepare the solution of (3.69b) which takes into account the periodic structure of the potential term $\det^{*,\sigma}/f_u^{*,\sigma}$.

First let us define the fundamental solutions on the unit periodic interval $(0, \omega)$, where $\omega = 2/n$ (see Fig. A.1). Let $Y_+(x)$ and $Y_-(x)$ denote the solutions of (3.69b) on $(0, \omega)$ satisfying

$$Y_+(0) = 1 = Y_-(\omega)$$

$$Y'_+(0) = 1 = Y'_-(\omega),$$ (2)

where $'$ denotes d/dx. It is clear that $Y_+(x)$ (resp. $Y_-(x)$) is strictly monotone increasing (resp. decreasing) and satisfies

$$Y_+(\omega - x) = Y_-(x)$$

$$Y'_+(\omega) = -Y'_-(0).$$ (3)

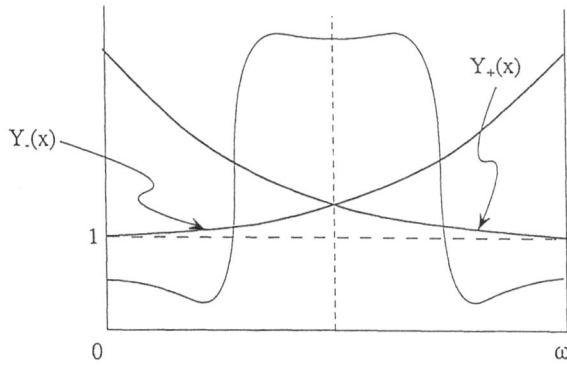

Figure A.1.

Lemma A.1. *There exist two linearly independent solution $F_\pm(x)$ of (3.69b) such that*

$$F_\pm(x) = e^{\pm \rho x} p_\pm(x),$$ (4)

where ρ is a real positive constant defined by

$$e^{\pm \rho \omega} = Y_+(\omega) \pm \left(Y_+(\omega)^2 - 1\right)^{\frac{1}{2}},$$ (5)

and $p_\pm(x)$ are ω-periodic C^1-functions satisfying

$$p_+(0) = 1 = p_-(0), \qquad p_+(\tfrac{\omega}{2}) = p_-(\tfrac{\omega}{2})$$ (6)

$$p'_+(0) + p'_-(0) = 0,$$ (7)

$$p'_+\left(\frac{\omega}{2}\right) + p'_-\left(\frac{\omega}{2}\right) = 0, \tag{8}$$

$$p_+(x_1^*)p_-(x_1^*) = p_+(x_2^*)p_-(x_2^*). \tag{9}$$

As a direct consequence of this lemma, we have

Corollary A.2.

$$F_\pm(0) = 1, \qquad F'_\pm(0) = \pm\rho + p'_\pm(0) \tag{i}$$

$$F'_+(0) + F'_-(0) = 0$$

$$h(x) = c_+^1 F_+(x) + c_-^1 F_-(x) \tag{ii}$$

$$k(x) = c_+^2 F_+(x) + c_-^2 F_-(x),$$

where

$$\mathbf{c}^1 = \begin{pmatrix} c_+^1 \\ c_-^1 \end{pmatrix} = \frac{1}{W_F} \begin{pmatrix} F'_-(0) \\ -F'_+(0) \end{pmatrix},$$

$$\mathbf{c}^2 = \begin{pmatrix} c_+^2 \\ c_-^2 \end{pmatrix} = \frac{1}{W_F} \begin{pmatrix} F'_-(1) \\ -F'_+(1) \end{pmatrix}.$$

Here W_F denotes the Wronskian of F_+ and F_-.

$$h^d(x) = d_+^1 F_+(x) + d_-^1 F_-(x) \tag{iii}$$

$$k^d(x) = d_+^2 F_+(x) + d_-^2 F_-(x),$$

where

$$\mathbf{d}^1 = \begin{pmatrix} d_+^1 \\ d_-^1 \end{pmatrix} = \frac{1}{W_F} \begin{pmatrix} -F_-(0) \\ F_+(0) \end{pmatrix},$$

$$\mathbf{d}^2 = \begin{pmatrix} d_+^2 \\ d_-^2 \end{pmatrix} = \frac{1}{W_F} \begin{pmatrix} -F_-(1) \\ F_+(1) \end{pmatrix}.$$

Proof of Lemma A.1.
We seek the solution of Floquet form (4) by using Y_+ and Y_-:

$$F(x) = c_+ Y_+ + c_- Y_-. \tag{10}$$

We determine the coefficients $c_\pm$ so that, after one period ω, F and F' are multiplied by some constant γ, namely

$$c_+Y_+(\omega) + c_-Y_-(\omega) = \gamma(c_+Y_+(0) + c_-Y_-(0))$$

$$c_+Y'_+(\omega) + c_-Y'_-(\omega) = \gamma(c_+Y'_+(0) + c_-Y'_-(0)).$$

(11)

Using (2) and (3), (11) becomes

$$\begin{pmatrix} Y_+(\omega) - \gamma & 1 - \gamma Y_+(\omega) \\ 1 & \gamma \end{pmatrix} \begin{pmatrix} c_+ \\ c_- \end{pmatrix} = \begin{pmatrix} 0 \\ 0 \end{pmatrix}.$$

(12)

Apparently (12) has nontrivial solutions if and only if γ is a root of

$$\gamma^2 - 2Y_+(\omega)\gamma + 1 = 0.$$

(13)

Since $Y_+(\omega) > 1$, (13) has two real positive distinct roots $\gamma_\pm$ such that

$$\gamma_\pm = Y_+(\omega) \pm \left(Y_+(\omega)^2 - 1\right)^{\frac{1}{2}}, \quad 0 < \gamma_- < 1 < \gamma_+.$$

We define ρ by $\rho \equiv (\log \gamma_+)/\omega$, then we have

$$e^{\pm\rho\omega} = \gamma_\pm, \qquad e^{\rho\omega} + e^{-\rho\omega} = 2Y_+(\omega).$$

(14)

Computing the eigenvectors $(c_+, c_-)^t$ associated with $\gamma_\pm$, the resulting solutions of the form (10) become

$$
\begin{aligned}
F_\pm(x) &\equiv \left(Y_+^2(\omega) - 1\right)^{-\frac{1}{2}} \{\pm e^{\pm\rho\omega}Y_+(x) \mp Y_-(x)\} \\
&= e^{\pm\rho x}p_\pm(x),
\end{aligned}
$$

(15a)

where $p_\pm(x)$ are defined by

$$p_\pm(x) \equiv \left(Y_+^2(\omega) - 1\right)^{-\frac{1}{2}} \{\pm e^{\pm\rho(\omega-x)}Y_+(x) \mp e^{\mp\rho x}Y_-(x)\}.$$

(15b)

It is easy to verify that $p_\pm(x)$ are ω-periodic functions of C^1-class. The properties (6) $\sim$ (9) are the direct consequences of the expression (15b) and the reflectional symmetry of Y_+ and Y_- at $x = \omega/2$, so the details are left to the reader. $\square$

Proof of Sublemma 2.
(a) Since $0 < h_i < h_j$ and $k_i > k_j > 0$, it is clear that $A_{ij} = h_ik_j - h_jk_i < 0$.

Properties (b) and (c) can be verified in a similar way with the aid of Lemma A.1 and Corollary A.2, so we prove only (b) in what follows.

(b) It follows from Corollary A.2 that h_i and k_j have expressions as

$$h_i = {}^t\mathbf{c}^1 \cdot \mathbf{F}_i, \qquad k_j = {}^t\mathbf{c}^2 \cdot \mathbf{F}_j,$$

where $\mathbf{c}^i = {}^t(c_+^i, c_-^i)$ and $\mathbf{F}_i = {}^t(F_+(x_i^*), F_-(x_i^*))$. We rewrite $A_{i+2,j+2}$ as

$$A_{i+2,j+2} = \begin{vmatrix} h_{i+2} & k_{i+2} \\ h_{j+2} & k_{j+2} \end{vmatrix} = \begin{vmatrix} {}^t\mathbf{c}^1 \cdot \mathbf{F}_{i+2} & {}^t\mathbf{c}^2 \cdot \mathbf{F}_{i+2} \\ {}^t\mathbf{c}^1 \cdot \mathbf{F}_{j+2} & {}^t\mathbf{c}^2 \cdot \mathbf{F}_{j+2} \end{vmatrix}. \tag{16}$$

On the other hand, we see from Lemma A.1 that

$$\mathbf{F}_{i+2} = T\mathbf{F}_i, \tag{17}$$

where $T = \begin{pmatrix} e^{\rho\omega} & 0 \\ 0 & e^{-\rho\omega} \end{pmatrix}$. Note that $|T| = 1$. Substituting (17) into (16), we have

$$
\begin{aligned}
A_{i+2,j+2} &= \begin{vmatrix} {}^t\mathbf{c}^1 T\mathbf{F}_i & {}^t\mathbf{c}^2 T\mathbf{F}_i \\ {}^t\mathbf{c}^1 T\mathbf{F}_j & {}^t\mathbf{c}^2 T\mathbf{F}_j \end{vmatrix} \\[2mm]
&= \left| \begin{pmatrix} {}^t\mathbf{c}^1 \\ {}^t\mathbf{c}^2 \end{pmatrix} T(\mathbf{F}_i\mathbf{F}_j) \right| \\[2mm]
&= \begin{vmatrix} {}^t\mathbf{c}^1 \cdot \mathbf{F}_i & {}^t\mathbf{c}^1 \cdot \mathbf{F}_j \\ {}^t\mathbf{c}^2 \cdot \mathbf{F}_i & {}^t\mathbf{c}^2 \cdot \mathbf{F}_j \end{vmatrix} \\[2mm]
&= A_{ij},
\end{aligned}
$$

which completes the proof. $\square$

Proof of Sublemma 3.
We shall prove only (a), since the remaining ones can be shown in a similar way. First Wronskians are given by

$$W(h,k) = \begin{vmatrix} h(0) & k(0) \\ h'(0) & k'(0) \end{vmatrix} = \begin{vmatrix} 1 & k(0) \\ 0 & k'(0) \end{vmatrix} = k'(0),$$

$$\tag{18}$$

$$W(h^d, k^d) = \begin{vmatrix} h^d(0) & k^d(0) \\ (h^d)'(0) & (k^d)'(0) \end{vmatrix} = \begin{vmatrix} 0 & k^d(0) \\ 1 & (k^d)'(0) \end{vmatrix} = -k^d(0).$$

In view of Corollary A.2, we see that

$$k'(0) = F_+'(0)(c_+^2 - c_-^2) = (\rho + p_+'(0))(c_+^2 - c_-^2)$$

$$\tag{19}$$

$$-k^d(0) = -(d_+^2 + d_-^2).$$

Using Corollary A.2, (18) and (19), we see that

$$
\begin{aligned}
I^N \equiv \ & \frac{A_{12}}{W(h, k)} = \frac{1}{F'_+(0)(c_+^2 - c_-^2)} \left[\{ c_+^1 F_+(x_1^*) + c_-^1 F_-(x_1^*) \} \right. \\
& \times\ \{ c_+^2 F_+(x_2^*) + c_-^2 F_-(x_2^*) \} - \{ c_+^1 F_+(x_2^*) + c_-^1 F_-(x_2^*) \} \\
& \times\ \{ c_+^2 F_+(x_1^*) + c_-^2 F_-(x_1^*) \}]
\end{aligned}
$$

$$
\begin{aligned}
I^D \equiv \ & \frac{A_{12}^d}{W(h^d, k^d)} = -\frac{1}{d_+^2 + d_-^2} \left[\{ d_+^1 F_+(x_1^*) + d_-^1 F_-(x_1^*) \} \right. \\
& \times\ \{ d_+^2 F_+(x_2^*) + d_-^2 F_-(x_2^*) \} - \{ d_+^1 F_+(x_2^*) + d_-^1 F_-(x_2^*) \} \\
& \times\ \{ d_+^2 F_+(x_1^*) + d_-^2 F_-(x_1^*) \}] \ .
\end{aligned}
$$

Substituting the expression of Corollary A.2 into the I^N and I^D, we obtain

$$
\begin{aligned}
I^N = \ & \frac{1}{(F'_-(1) + F'_+(1))W_F} \left[-\{ F_+(x_1^*) + F_-(x_1^*) \} \right. \\
& \times\ \{ F'_-(1)F_+(x_2^*) - F'_+(1)F_-(x_2^*) \} + \{ F_+(x_2^*) + F_-(x_2^*) \} \\
& \times\ \{ F'_-(1)F_+(x_1^*) - F'_+(1)F_-(x_1^*) \}] \ ,
\end{aligned}
\tag{20a}
$$

$$
\begin{aligned}
I^D = \ & \frac{1}{(F_+(1) - F_-(1))W_F} \left[\{ -F_+(x_1^*) + F_-(x_1^*) \} \right. \\
& \times\ \{ -F_-(1)F_+(x_2^*) + F_+(1)F_-(x_2^*) \} - \{ -F_+(x_2^*) + F_-(x_2^*) \} \\
& \times\ \{ -F_-(1)F_+(x_1^*) + F_+(1)F_-(x_1^*) \}] \ .
\end{aligned}
\tag{20b}
$$

Next we shall express $F_\pm(1)$ and $F'_\pm(1)$ in terms of $p_+(1)$ and $p'_+(1)$. Since $\omega n/2 = 1$, we see from (4) that

$$
F_\pm(1) = e^{\pm \rho \omega n/2} p_\pm(1)
$$
$$
F'_\pm(1) = \pm \rho e^{\pm \rho \omega n/2} p_\pm(1) + e^{\pm \rho \omega n/2} p'_\pm(1).
\tag{21}
$$

Recalling the ω-periodicity of $p_\pm(x)$, we have from (6) $\sim$ (8) that

$$
p_+(1) = p_-(1) \quad \text{and} \quad p'_+(1) + p'_-(1) = 0.
\tag{22}
$$

Substituting (22) into (21), we have

$$
F_\pm(1) = e^{\pm \rho \omega n/2} p_+(1)
$$
$$
F'_\pm(1) = \pm (\rho p_+(1) + p'_+(1)) e^{\pm \rho \omega n/2}.
\tag{23}
$$

Inserting (23) into (20), we obtain

$$I^N = \frac{1}{EW_F}\left[\{F_+(x_1^*) + F_-(x_1^*)\}\{e^{-\rho\omega n/2}F_+(x_2^*) + e^{\rho\omega n/2}F_-(x_2^*)\}\right.$$

$$\left. -\{F_+(x_2^*) + F_-(x_2^*)\}\{e^{-\rho\omega n/2}F_+(x_1^*) + e^{\rho\omega n/2}F_-(x_1^*)\}\right],$$

$$I^D = \frac{1}{EW_F}\left[\{F_+(x_1^*) - F_-(x_1^*)\}\{-e^{-\rho\omega n/2}F_+(x_2^*) + e^{\rho\omega n/2}F_-(x_2^*)\}\right.$$

$$\left. -\{F_+(x_2^*) - F_-(x_2^*)\}\{-e^{-\rho\omega n/2}F_+(x_1^*) + e^{\rho\omega n/2}F_-(x_1^*)\}\right],$$

where $E \equiv e^{\rho\omega n/2} - e^{-\rho\omega n/2}$. Expanding these expressions, we easily see that both I^N and I^D are equal to $\{F_+(x_1^*)F_-(x_2^*) - F_-(x_1^*)F_+(x_2^*)\}/W_F$, which completes the proof. □

Proof of Sublemma 4.

We shall prove only (a). First noting that $h(x)$ is equal to the fundamental solution $Y_+(x)$ on $(0, \omega)$ (see (2)), we can write

$$\frac{h_2}{h_1} = \frac{h(x_2^*)}{h(x_1^*)} = \frac{Y_+(x_2^*)}{Y_+(x_1^*)}. \tag{24}$$

We set for simplicity

$$\alpha \equiv Y_+(x_1^*), \qquad \beta \equiv Y_+(x_2^*), \qquad \Omega \equiv Y_+(\omega). \tag{25}$$

Since Y_+ is strictly monotone increasing, it is obvious that

$$0 < \alpha < \beta < \Omega.$$

By a simple computation, we easily see that $h^d(x)$ can be written as

$$h^d(x) = \frac{1}{Y_-'(0)}\{-Y_-(0)Y_+(x) + Y_-(x)\}. \tag{26}$$

Using this expression, we have

$$\frac{h_2^d}{h_1^d} = \frac{h^d(x_2^*)}{h^d(x_1^*)} = \frac{-Y_-(0)Y_+(x_2^*) + Y_-(x_2^*)}{-Y_-(0)Y_+(x_1^*) + Y_-(x_1^*)}. \tag{27}$$

Because of symmetry of Y_+ and Y_- (see Figure A.1), we have $Y_-(x_1^*) = \beta$, $Y_-(x_2^*) = \alpha$, and $Y_-(0) = \Omega$. Hence (27) becomes

$$\frac{h_2^d}{h_1^d} = \frac{\alpha - \Omega\beta}{\beta - \Omega\alpha}.$$

On the other hand, it is clear from (24) and (25) that $h_2/h_1 = \beta/\alpha$. Thus we have

$$\frac{h_2^d}{h_1^d} - \frac{h_2}{h_1} = \frac{\alpha^2 - \beta^2}{\alpha(\beta - \alpha\Omega)}.$$

Noting that $\beta - \alpha\Omega = Y_-'(0)h^d(x_1^*) < 0$ (see (26)), this implies the conclusion. □

References

[1] J.Alexander, R.Gardner and C.Jones, A topological invariant arising in the stability analysis of travelling waves, J. reine angew. Math., **410** (1990), 167-212.

[2] N.D.Alikakos, P.W.Bates, and G.Fusco, Slow motion for the Cahn- Hilliard equation in one space dimension, J. Differential Equations, **90** (1991), 81-135.

[3] L.Bronsard and R.V.Kohn, On the slowness of the phase boundary motion in one space dimension, Comm. Pure Appl. Math., **43** (1990), 983-997.

[4] L.Bronsard and R.V.Kohn, Motion by mean curvature as the singular limit of Ginzburg-Landau dynamics, J. Diff., Eqns., **90** (1991) 211-237.

[5] P.Brunovský and B.Fiedler, Connection orbits in scalar reaction diffusion equations, Dynamics Reported vol 1 (1988), 57-90, Wiley.

[6] G.Caginalp, An analysis of a phase field model of a free boundary, Arch. Rat. Mech. Anal., **92** (1986), 205-245.

[7] G.Caginalp and Y.Nishiura, The existence of travelling waves for phase field equations and convergence to sharp interface models in the singular limit, Quarterly of Appl. Math., **49** (1) (1991), 147-162.

[8] J.Carr, M.E.Gurtin, and M.Slemrod, Structured phase transitions on a finite interval, Arch. Rat. Mech. Anal., **86** (1984), 317-351.

[9] J.Carr and R.L.Pego, Metastable patterns in solutions of $u_t = \varepsilon^2 u_{xx} - f(u)$, Comm. Pure Appl. Math., **42** (1989), 523-576.

[10] R.G.Casten and C.J.Holland, Instability results for reaction diffusion equations with Neumann boundary conditions, J. Diff. Eqns., **27** (1978), 266-273.

[11] X.-Y.Chen, Dynamics of interfaces in reaction diffusion systems, Hiroshima Math. J., **21** (1991), 47-83.

[12] Y.-G.Chen, Y.Giga, and S.Goto, Uniqueness and existence of viscosity solutions of generalized mean curvature flow equations, J. Differential Geometry 33 (1991), 749-786.

[13] S.-N.Chow, B.Deng, and D.Terman, The bifurcation of homoclinic and periodic orbits from two heteroclinic orbits, SIAM J. Math. Anal., **21** (1990), 179-204.

[14] B.Deng, The bifurcations of countable connections from a twisted heteroclinic loop, SIAM J. Math. Anal. **22** (1991), 653-679.

[15] L.C.Evans and J.Spruck, Motion of level sets by mean curvature *I*, J. Differential Geometry 33 (1991), 601-633.

[16] P.C.Fife, Boundary and interior transition layer phenomena for pairs of second order differential equations, J. Math. Anal., **54** (1976), 497-521.

[17] P.C.Fife, Propagator-controller systems and chemical patterns, in *Nonequilibrium Dynamics in Chemical Systems*, A.Pacault and C.Vidal, eds., Springer-Verlag (1984), 76-88.

[18] P.C.Fife, Understanding the patterns in the BZ reagent, J. Statist. Phys., **39** (1985), 687-703.

[19] P.C.Fife, *Dynamics of Internal Layers and Diffusive Interfaces*, CBMS-NSF Regional Conference Series in Applied Mathematics #**53**, SIAM, Philadelphia (1988).

[20] P.C.Fife, Pattern dynamics for parabolic PDE's, to appear in the Proc. of IMA Conference "Introduction to dynamical systems", Sept. 1989.

[21] P.C.Fife and W.M.Greenlee, Interior transition layers for elliptic boundary value problems with a small parameter, Russian Math. Surveys, **29**:4 (1974), 103-131.

[22] P.C.Fife and J.B.McLeod, The approach of solutions of nonlinear diffusion equations to travelling front solutions, Arch. Rat. Mech. Anal., **65** (1977), 335-361.

[23] H.Fujii and Y.Hosono, Neumann layer phenomena in nonlinear diffusion systems, *Recent Topics in Nonlinear PDE*, M.Mimura and T.Nishida, eds. Math. Studies, **98**, North-Holland, Amsterdam (1983), 21-38.

[24] H.Fujii, M.Mimura and Y.Nishiura, A picture of the global bifurcation diagram in ecological interacting and diffusing systems, Phisica **5D** (1982), 1-42.

[25] H.Fujii, Y.Nishiura and Y.Hosono, On the structure of multuple existence of stable stationary solutions in systems of reaction- diffusion equations, Patterns and Waves - Qualitative Analysis of Nonlinear Differential Equations, T.Nishida, M.Mimura and H.Fujii, eds., Stud. Math. Appl., **18** (1986), 157-219.

[26] G.Fusco and J.K.Hale, Slow-motion manifolds, dormant Instability, and singular perturbations, J. Dynamics and Diff. Eqns., **1** (1989), 75-94.

[27] R.Gardner and C.K.R.T.Jones, A stability index for steady state solutions of boundary value problems for parabolic systems, J. Diff. Eqns., **91** (1991), 181-203.

[28] J.K.Hale and K.Sakamoto, Existence and stability of transition layers, Japan J. Appl. Math., **5** (1988), 367-405.

[29] D.Henry, *Geometric theory of semilinear parabolic equations*, Lecture Notes in Mathematics **840**, Springer-Verlag, Berlin, New York, 1981.

[30] D.Hilhorst, Y.Nishiura, and M.Mimura, A free boundary problem arising in some reacting-diffusing system, Proc. Roy. Soc. Edinburgh, **118A** (1991), 355-378.

[31] T.Ikeda and Y.Nishiura, Pattern selection for two breathers, SIAM Appl. Math., in press.

[32] M.Ito, A remark on singular perturbation methods, Hiroshima Math. J., **14** (1985), 619-629.

[33] J.P.Keener and J.J.Tyson, Spiral waves in the Belousov- Zhabotinsky reaction, Physica, **21D** (1986), 307-324.

[34] H.Kokubu, Homoclinic and heteroclinic bifurcation of vector fields, Japan J. Appl. Math., **5** (1987), 455-501.

[35] H.Kokubu, Y.Nishiura, and H.Oka, Heteroclinic and homoclinic bifurcation in bistable reaction diffusion systems, J. Diff. Eqns., **86** (2) (1990), 260-341.

[36] J.S.Langer, Instabilities and pattern formation in crystal growth, Review of Modern Physics, **52** (1) (1980), 1-28.

[37] H.Matano, Asymptotic behavior and stability of solutions of semilinear diffusion equations, Publ. RIMS, Kyoto Univ., **15** (1979), 401-454.

[38] H.Meinhardt, *Models of biological pattern formation*. Academic Press, 1982.

[39] M.Mimura, M.Tabata, and Y.Hosono, Multiple solutions of two-point boundary value problems of Neumann type with a small parameter, SIAM J. Math. Anal., **11** (1980), 613-631.

[40] J.D.Murray, *Mathematical Biology*. Springer-Verlag, 1989.

[41] Y.Nishiura, Global structure of bifurcating solutions of some reaction-diffusion systems, SIAM J. Math. Anal., **13** (1982), 555-593.

[42] Y.Nishiura, Singular limit approach to stability and bifurcation for bistable reaction diffusion systems. In "Proceeding of the Workshop on Nonlinear PDE's, Provo, Utah, March 1987" (P.Bates and P.Fife, eds.), Rocky Mountain J. Math., **21** (2) (1991), 727-767.

[43] Y.Nishiura, Singular limit eigenvalue problem in higher dimensional space, Proceeding of International Symposium on Functional Differential Equations and Related Topics, Kyoto, Japan, 30 August – 2 September, T.Yoshizawa and J.Kato, eds., World Scientific, 1991.

[44] Y.Nishiura and H.Fujii, Stability of singularly perturbed solutions to systems of reaction diffusion equations, SIAM J. Math. Anal., 18 (1987), 1726-1770.

[45] Y.Nishiura and H.Fujii, SLEP method to the stability of singularly perturbed solutions with multiple internal transition layers in reaction-diffusion systems, Proc. NATO Workshop "Dynamics of Infinite Dimensional Systems", J. Hale and S.N. Chow, eds., NATO ASI Series F37 (1986), 211-230.

[46] Y.Nishiura and M.Mimura, Layer oscillations in reaction- diffusion systems, SIAM J. Appl. Math., 49 (1989), 481-514.

[47] Y.Nishiura, M.Mimura, H.Ikeda, and H.Fujii, Singilar limit analysis of stability of travelling wave solutions in bistable reaction-diffusion systems, SIAM J. Math. Anal., 21 (1990), 85- 122.

[48] Y.Nishiura and H.Suzuki, Coalescence,repulsion, and nonuniqueness for singular limit reaction diffusion systems, to appear in the Proceedings of the International Taniguchi Symposium on Nonlinear Partial Differential Equations and Applications, Katata, 23 August – 29 August, 1989 (Eds. T.Nishida and M.Mimura).

[49] Y.Nishiura and H.Suzuki, Stability of generic interface of reaction diffusion systems in higher space dimensions, in preparation.

[50] Y.Nishiura and T.Tsujikawa, Instability of singularly perturbed Neumann layer solutions in reaction-diffusion systems, Hiroshima Math. J., 20 (1990), 297-329.

[51] T.Ohta and M.Mimura, Pattern dynamics in excitable reaction- diffusion media, Formation, Dynamics and Stabilities of Patterns, K.Kawasaki, M.Suzuki, and A.Onuki, eds., 1, World Scientific, 1990.

[52] T.Ohta, M.Mimura, and R.Kobayashi, Higher-dimensional localized pattern in excitable media, Physics 34D (1989), 115-144.

[53] P.Pelcé (editor), Dynamics of curved fronts, Perspective in Physics, Academic Press, 1988.

[54] J.Rinzel and D.Terman, Propagation phenomena in a bistable reaction-diffusion systems, SIAM J. Appl. Math., 42 (1982) 1111-1137.

[55] K.Sakamoto, Construction and stability analysis of transition layer solutions in reaction-diffusion systems, Tôhoku Math. J., 42 (1990), 17-44.

[56] H.Suzuki, Y.Nishiura, and H.Ikeda, Stability of traveling waves and a relation between the Evans function and the SLEP equation, submitted for publication.

[57] M.Taniguchi and Y.Nishiura, Instability of planar interfaces in reaction diffusion systems, SIAM Math. Anal., in press.

[58] A.M.Turing, The chemical basis of morphogenesis, Phil. Trans. Roy. Soc. Lond., B237 (1952), 37-72.

[59] R.S.Varga, Matrix iterative analysis, Prentice-Hall, 1962.

[60] J.H.Wilkinson, The algebraic eigenvalue problem, Oxford, Clarendon Press, 1965.

Recent Advances in Regularity of Second-order Hyperbolic Mixed Problems, and Applications*

I. Lasiecka and R. Triggiani

Department of Applied Mathematics, Thornton Hall, University of Virginia, Charlottesville, VA, 22903 USA

1. Introduction

The present paper centers on second-order hyperbolic equations in the unknown $w(t, x)$:

$$w_{tt} + A(x, \partial)w = f \quad \text{in } Q = (0, T] \times \Omega \tag{1.1}$$

augmented by initial conditions

$$w(0, \cdot) = w_0; \quad w_t(0, \cdot) = w_1 \quad \text{in } \Omega \tag{1.2}$$

and suitable boundary conditions either of Dirichlet type

$$w|_\Sigma = u \quad \text{in } \Sigma = (0, T] \times \Gamma, \tag{1.3D}$$

or else of Neumann type

$$\frac{\partial w}{\partial \nu_A} = u \quad \text{in } \Sigma, \tag{1.3N}$$

where $\partial/\partial\nu_A$ denotes the corresponding co-normal derivative. Here and throughout, Ω is a general open bounded domain in R^n, n typically ≥ 2, with boundary $\partial\Omega = \Gamma$ assumed 'smooth' (the 'degree' of smoothness depending on the 'degree' of regularity of the solutions we wish to consider). Moreover, $A(x, \partial)$ denotes a second-order elliptic operator on Ω:

$$
\begin{cases}
A(x, \partial) = -\sum_{i,j}^n \dfrac{\partial}{\partial x_i}\left(a_{ij}(x)\dfrac{\partial}{\partial x_j}\right), & (1.4a) \\[2ex]
\sum_{i,j}^n a_{ij}(x)\xi_i\xi_j \geq c \sum_i^n \xi_i^2 \quad \text{constant } c > 0, & (1.4b)
\end{cases}
$$

with suitably smooth coefficients $a_{ij}(x) = a_{ji}(x)$.

The main goal of this paper is twofold: (i) to present (a selected sample of) regularity results for the solutions, and traces thereof, of the mixed problems (1.1), (1.2), (1.3D),

---

* Submitted August 23, 1990. Research partially supported by the National Science Foundation under Grant DMS-8902811

or (1.3N), which have become available over the past years; (ii) to demonstrate the applicability of these results to (a selected sample of) semi-linear mixed problems and boundary control problems, where the optimal/sharp linear theory in (i) plays a critical role. A subordinate goal is to show the usefulness in the analysis of novel abstract operator models to describe the corresponding partial differential equation mixed problems, both in the solution of specific questions as well as in providing a unified framework for the class of hyperbolic problems (and, in fact, of plate problems as well). But there is no space in this paper to even touch the corresponding advances in regularity theory of plate mixed problems and related applications, which have followed over the past few years under the stimulus of ideas and techniques successfully employed for second-order hyperbolic equations).

The regularity theory of (linear) parabolic mixed problems has been in essentially optimal shape for quite some time, e.g., [L-M.1] within the Hilbert space framework. Later, a purely operator approach was given yielding an explicit representation formula in the general case where the free problem generates an analytic semigroup (see Notes of Section 4). This representation — when merged with elliptic theory and Sobolev theory — permits then to re-obtain the preceding theory (with some improvements) and allows the flexibility to work outside the L_2-framework, e.g., [Las. 1]: optimal regularity results in a context different from L_2, e.g., [L-T.9], [L-T.10], play a crucial role, e.g., in the Riccati theory (differential as well as algebraic) associated with parabolic problems.

The situation has been quite different in the literature in the case of the hyperbolic mixed problems (1.1)–(1.3) (and plate problems as well), except for the one-dimensional case, where elementary methods and explicit formulas provide the optimal regularity results. The present theory in higher dimension is the result of the past recent years. As it turns out, the situation is far simpler in the Dirichlet case (1.3D), where a regularity theory has emerged, which is optimal and which does not depend on the dimension of Ω. Moreover, the key building block of this regularity theory can be given through a tricky but now well understood energy method within the framework of differential calculus (see Notes of Section 2). In contrast, the situation is drastically different in the Neumann case (1.3N). First, the optimal regularity theory of one dimension is patently false, by 'much', in higher dimension. Second, the key building blocks of the presently available sharp regularity theory are obtained by far more complicated energy methods within the frameworks of pseudo-differential calculus in the general case. Special geometries (spheres, parallelepipeds, ...) can be handled through technical, hard-analysis computations based on appropriate eigenfunction expansions to yield higher regularity properties. Thus, in the Neumann case, the regularity theory appears to depend on the geometry of Ω for $\dim \Omega \geq 2$ (see Notes of Section 3).

The paper is divided in two parts. Part I presents some highlights of the regularity theory in the Dirichlet case (Section 2) and in the Neumann case (Section 3). Operator models in differential or integral form providing explicit representation formulas on the data are introduced in Section 4, and the regularity theory of Sections 2 and 3 is then recast within the operator representation of the mixed problems.

Part II deals with several applications: well-posedness of the semi-linear Neumann problem (Section 5); local exponential stability of semi-linear problems (Section 6); exact controllability of semi-linear problems (Section 7); and, finally, Riccati differential equations (quadratic) and hyperbolic mixed problems (Section 8). In each case, the role of the associated linear theory is of paramount importance and duly stressed.

With such a scope, there is no space for inclusion of proofs of all results presented. Instead, we have opted for, and consistently followed, the following strategy. The regularity results are simply listed and referred to the literature. Instead, in Part II the paper provides a sufficiently detailed and conceptual scheme of the proofs which in particular includes all those parts where the regularity theory of Part I plays a critical role; here we care to emphasize the technical points, where the regularity of part I makes the proof succeed, while the earlier regularity theory would instead be insufficient and inadequate.

Part I: Regularity Theory

2. Regularity under Dirichlet Boundary Conditions

Throughout this section we deal with the following mixed problem for second-order hyperbolic equations of Dirichlet type:

$$\begin{cases} w_{tt} + A(x, \partial)w = F & \text{in } Q = (0, T] \times \Omega, & (2.1\text{a}) \\ w(0, \cdot) = w_0 : w_t(o, \cdot) = w_1 & \text{in } \Omega, & (2.1\text{b}) \\ w|_\Sigma = u & \text{in } \Sigma = (0, T] \times \Gamma, & (2.1\text{c}) \end{cases}$$

mentioned in the Introduction, where $A(x, \partial)$ is defined in (1.4). The regularity question consists of a quantitative description of the map

$$\{w_0, w_1, f, u\} \rightarrow \left\{ w, w_t, \cdots, \frac{\partial w}{\partial v} \right\} \tag{2.2}$$

from the data to the solution (and its Neumann trace) of problem (2.1). Below we shall collect only some distinctive results of the map (2.2), including those that are relevant to the subsequent sections. For additional results and proofs we refer to [L-L-T.1] and [L-T.8, Sect. 3], which present a rather comprehensive study following the initial contributions [L-T.1], [L-T.2], [Lio.2], in the case of $u \in L_2(\Sigma)$; see also *Notes* at the end of this section.

Theorem 2.1. With reference to (2.1), assume

$$f \in L_1(0, T; H^{-1}(\Omega)), \tag{2.3}$$
$$w_0 \in L_2(\Omega); \quad w_1 \in H^{-1}(\Omega), \tag{2.4}$$
$$u \in L_2(0, T; L_2(\Gamma)) \equiv L_2(\Sigma) \tag{2.5}$$

(without compatibility conditions). Then, the unique solution of (2.1) satisfies the interior regularity

$$w \in C([0, T]; L_2(\Omega)), \tag{2.6}$$
$$w_t \in C([0, T]; H^{-1}(\Omega)), \tag{2.7}$$

and the boundary regularity

$$\frac{\partial w}{\partial \nu} \in H^{-1}(\Sigma). \tag{2.8}$$

Theorem 2.2.
(a) (interior regularity). With reference to (2.1), assume

$$f \in L_1(0, T; L_2(\Omega)), \tag{2.9}$$
$$w_0 \in H^1(\Omega); \quad W_1 \in L_2(\Omega), \tag{2.10}$$
$$u \in C([0, T]; H^{\frac{1}{2}}(\Gamma)) \cap H^1(0, T; L_2(\Gamma)), \tag{2.11}$$

along with the compatibility condition

$$w_0|_\Gamma = u|_{t=0} \in H^{1/2}(\Gamma). \tag{2.12}$$

Then the unique solution of (2.1) satisfies

$$w \in C([0, T]; H^1(\Omega)); \tag{2.13}$$
$$w_t \in C([0, T]; L_2(\Omega)). \tag{2.14}$$

(b) (boundary regularity). In addition to (2.9), (2.10), (2.12), assume now

$$u \in H^1(\Sigma) \tag{2.15}$$

stronger than (2.11). Then, the unique solution of (2.1) satisfies

$$\frac{\partial w}{\partial \nu} \in L_2(\Sigma). \tag{2.16}$$

Theorem 2.3.
(a) (interior regularity). With reference to (2.1), assume

$$f \in L_1(0, T; H^1(\Omega)), \tag{2.17}$$
$$w_0 \in H^2(\Omega); \quad w_1 \in H^1(\Omega), \tag{2.18}$$
$$u \in C([0, T]; H^{\frac{3}{2}}(\Gamma)) \cap H^2(0, T; L_2(\Gamma)), \tag{2.19a}$$
$$u_t \in C([0, T]; H^{\frac{1}{2}}(\Gamma)), \tag{2.19b}$$

along with the compatibility conditions

$$w_0|_\Gamma = u|_{t=0} \in H^{\frac{3}{2}}(\Gamma); \qquad w_1|_\Gamma = u_t|_{t=0} \in H^{1/2}(\Gamma). \tag{2.20}$$

Then, the unique solution of (2.1) satisfies

$$w \in C([0, T]; H^2(\Omega)); \tag{2.21}$$
$$w_t \in C([0, T]; H^1(\Omega)). \tag{2.22}$$

(b) (boundary regularity). In addition to (2.17), (2.18), (2.20), assume now

$$u \in H^2(\Sigma) \tag{2.23}$$

stronger than (2.19). Then, the unique solution of (2.1) satisfies

$$\frac{\partial w}{\partial \nu} \in H^1(\Sigma). \tag{2.24}$$

Remarks 2.1.

(i)   In Theorem 2.1, the boundary regularity (2.8) does NOT follow from the interior regularity results (2.6), (2.7): it is an independent regularity result.

(ii)   Let $u \in H^1(\Sigma)$. Then *a fortiori* [L-M.1, Thm. 3.1, p. 19],

$$u \in C([0, T]; [H^1(\Gamma), H^0(\Gamma)]_{\frac{1}{2}}) = C([0, T]; H^{\frac{1}{2}}(\Gamma)).$$

Thus, assumption (2.15) is stronger than assumption (2.11). In Theorem 2.2, even under the stronger assumption (2.15), the boundary regularity (2.16) does NOT follow from the interior regularity (2.13), (2.14): it is an independent regularity result.

(iii) Let $u \in H^2(\Sigma)$. Then, *a fortiori* [L-M.1, Thm. 3.1, p. 19],

$$u \in C([0, T]; [H^2(\Gamma), H^0(\Gamma)]_{\frac{1}{4}}) = C([0, T]; H^{\frac{3}{2}}(\Gamma));$$

$$u_t \in C([0, T]; [H^2(\Gamma), H^0(\Gamma)]_{\frac{3}{4}}) = C([0, T]; H^{\frac{1}{2}}(\Gamma)).$$

Thus, assumption (2.23) is stronger than assumptions (2.19a–b). In Theorem 2.3, even under the stronger assumption (2.23), the boundary regularity (2.24) does NOT follow from the interior regularity (2.21), (2.22).

(iv)   In Theorem 2.3, assumption (2.19b) is needed only for conclusion (2.22), not for conclusion (2.21).

(v)   Theorem 2.2 with $u \in H^1(\Sigma)$ is 'almost' an interpolation result between Theorem 2.1 and Theorem 2.3 with $u \in H^2(\Sigma)$. For interpolation results, we refer to [L-L-T.1, Section 2.7, p. 164].

(vi)   The conclusion (2.6) of Theorem 2.1 with say $w_0 = w_1 = 0$, $f = 0$ should be contrasted with well-known corresponding results for second-order *parabolic* equations, where with Dirichlet datum $u \in L_2(\Sigma)$ and, say, homogeneous other data, the corresponding solution may fail to be in $L_2(\Omega)$ at a pre-assigned time t, even in the one-dimensional case [Lio.1, p. 202].

(vii) The above regularity results for w, w_t of (2.1) is optimal, in the sense that it coincides with that available in the one-dimensional case ($\dim \Omega = 1$), where one can write an explicit formula for the solution w (e.g., [L-T.1, Example 5.1, p. 52]). More importantly, the aforementioned optimality of the regularity of w, w_t stems also from more recent 'exact boundary controllability' results for (2.1), which provide, in particular, surjectivity of the map, say, $u \to \{w, w_t\} : L_2(\Sigma) \to L_2(\Omega) \times H^{-1}(\Omega)$; or $H_0^1(0, T; L_2(\Gamma)) \to H_0^1(\Omega) \times L_2(\Omega)$ [H.1], [L-T.17, App.], [Lio.3], [Tr.5].

(viii) Other regularity results are given in [L-L-T.1] and [L-T.8, Sect. 3]. They include: additional results where hypotheses made on u are not symmetric with respect to time and space; results where u is 'less smooth' than $L_2(\Sigma)$, in particular, $u(x, t) = h(t)\delta_b(x)$, with $h \in L_2(0, T)$ and $\delta_b(x) = $ Dirac mass concentrated at b, $b \in \Gamma$. For instance, one result to be invoked in Section 7 is as follows:

$$f \equiv 0, \qquad w_0 = w_1 = 0,$$

$$u \in H_0^1(0, T; L_2(\Gamma)) \to \{w, w_t\} \in C([0, T]; H^{\frac{1}{2}}(\Omega) \times L_2(\Omega)), \qquad (2.25)$$

compare with assumption (2.11) and conclusion (2.13), (2.14).

(ix)   Theorem 2.1, Theorem 2.2 with $u \in H^1(\Sigma)$, and Theorem 2.3 with $u \in H^2(\Sigma)$
       extend to the case where the coefficients of the uniformly elliptic operator $A(\ ,\ )$
       in (1.4) may depend on t: $a_{ij}(x, t)$ provided that they are smooth enough. Example:
       Theorem 2.2 extends to the present case provided that $a_{ij}(x, t) \in W^{1,\infty}(Q)$:
       $\frac{\partial a_{ij}}{\partial t}$, $\frac{\partial a_{ij}}{\partial x_k} \in L_\infty(Q)$; see [L-L-T.1, Section 4], [Lio.2].

Notes. *The basic interior regularity result* (2.6), (2.7) *of Theorem* 1.1:

$$f = 0; \quad w_0 = w_1 = 0; \quad u \in L_2(\Sigma) \to \{w, w_t\} \in C([0, T]; L_2(\Omega) \times H^{-1}(\Omega)), \qquad (2.26)$$

as well as the related trace regularity (2.16) *of Theorem* 2.2 *with* $u \equiv 0$:

$$\{w_0, w_1, f\} \in H_0^1(\Omega) \times L_2(\Omega) \times L_1(0, T; L_2(\Omega)) \to \frac{\partial w}{\partial \nu} \in L_2(\Sigma) \qquad (2.27)$$

were first obtained in [L-T.1], [L-T.2] *by pseudo-differential/operator methods and later in* [Lio.2] *by energy methods. All other results, which are based on these two, are taken from the subsequent paper* [L-L-T.1], *which along with* [L-T.8, Sect. 3] *contains further results. Prior literature on this subject includes* [L-M.1, vols. I&II], [Lio.1] — *which give the most updated account available in the late sixties* — *and* [Sa.1], [Sa.2]. *As a comparison, the regularity result* (2.26) *improves substantially upon the results of* [L-M.1, vol. II, p. 121]. *References* [Sa.1], [Sa.2] *study the regularity of the hyperbolic mixed problem* (2.1) *(and of higher order as well) with* $u \in H^k(\Sigma)$, $k \geq 1$ *on the half-space, by means of pseudo-differential operator techniques, and obtain results which are contained in Theorems* 2.2 *and* 2.3. *However, Theorems* 2.2 *and* 2.3 *are more precise, as they show that a regularity of* u *weaker than* $H^k(\Sigma)$, $k \geq 1$, *is sufficient to obtain the desired interior regularity of* w, w_t *on* Q, *while the full strength of the assumption* $u \in H^k(\Sigma)$ *is used to obtain the desired boundary regularity for* $\frac{\partial w}{\partial \nu}$ *on* Σ. *The original contributions* [L-T.1], [L-T.2] *showed first the interior regularity* (2.26), *by passing through the intermediate step of establishing first* $\{w, w_t\} \in L_2(0, T; L_2(\Omega) \times H^{-1}(\Omega))$ *by means of a pseudo-differential approach on the half space.*

   Next, this $L_2(0, T; L_2(\Omega) \times H^{-1}(\Omega))$ *interior regularity is then shown to be equivalent to the trace regularity* (2.27) *and finally boosted to* $C([0, T]; L_2(\Omega) \times H^{-1}(\Omega))$ *as in* (2.26), *through a general operator-theoretic approach (see also* [L-T.7]). *Instead,* [Lio.2] *and then* [L-L-T.1] *follow a reverse approach, that turns out to be simpler, to first show the trace regularity* (2.27) *and then the interior regularity* (2.26) *by duality or transposition. Thus, the key building block of the regularity theory is the trace regularity* (2.27) *(for which see Remark* 2.1(ii)). *The most direct way to obtain* (2.27) *is to multiply the homogeneous problem with* $u \equiv 0$ *by the multiplier* $h \cdot \nabla w$ *(going back, apparently to Rellich for elliptic problems), where* h *is a smooth vector field such that* $h = \nu$ *on* Γ. *After various integrations by parts, one obtains an identity which gives* $|\frac{\partial w}{\partial \nu}|^2_{L_2(\Sigma)}$ *in terms of the data* $\{w_0, w_1, f\}$ *in* $H_0^1(\Omega) \times L_2(\Omega) \times L_1(0, T; L_2(\Omega))$. *This technique has been successfully used also for exact controllability problems with* $u \in L_2(\Sigma_1)$, $u = 0$ *in* $\Sigma \setminus \Sigma_1$, *where the key equivalent goal is to show the reverse inequality for* T *sufficiently large:*

$$\int_{\Sigma_1} \left(\frac{\partial w}{\partial \nu}\right)^2 d\Sigma \geq c_T |\{w_0, w_1\}|^2_{H_0^1(\Omega) \times L_2(\Omega)}$$

for w *of* (2.1) *with* $f = 0$, $u = 0$ [L-T.22], [L-T.23], [H.1], [Lio.3], [Tr.5], [B-L-R], [Ta.1].

3. Regularity under Neumann Boundary Conditions

Throughout this section we take $\dim \Omega \geq 2$, and we deal with the following mixed problem for second-order hyperbolic equations of Neumann type

$$\begin{cases} w_{tt} + A(x, \partial)w = f & \text{in } Q = (0, T] \times \Omega, & (3.1a) \\ w(0, \cdot) = w_0 : \ w_t(0, \cdot) = w_1 & \text{in } \Omega & (3.1b) \\ \frac{\partial w}{\partial \nu_A} = U & \text{in } \Sigma = (0, T] \times \Gamma, & (3.1c) \end{cases}$$

mentioned in the Introduction with $A(x, \partial)$ defined in (1.4) and $\partial/\partial \nu_A$ the corresponding co-normal derivative (= normal derivative of $A(x, \partial) = -\Delta$). The following sharp regularity results for problem (3.1) are taken from the recent papers [L-T.3], [L-T.5] whose main results were announced in [Las.3], [L-T.4], [Tr.6]. First, we introduce the parameters α and β which, throughout this section, will take on only the following values for the following specified cases, where $\varepsilon > 0$ arbitrary:

$$\left. \begin{array}{l} \alpha = \frac{3}{5} - \varepsilon \\ \beta = \frac{3}{5} \end{array} \right\} : \quad \begin{array}{l} \text{for a general smooth, bounded domain } \Omega \\ \text{and a general operator } A(x, \partial) \text{ as in (1.4)} \end{array}$$

$$\alpha = \beta = \frac{2}{3} : \quad \text{for a sphere } \Omega \text{ and the Laplacian } -A(x, \partial) = \Delta,$$

$$\alpha = \beta = \frac{3}{4} : \quad \text{for a parallelepiped } \Omega \text{ and the}$$
$$\text{Laplacian } -A(x, \partial) = \Delta. \qquad (3.2c)$$

Throughout the results listed below we always take $\dim \Omega \geq 2$, unless otherwise noted; see Remark 3.1(iv) for $\dim \Omega = 1$.

Theorem 3.1. ([L-T.3], [L-T.5, Thm. A]) With reference to (3.1), assume

$$w_0 = w_1 = 0; \quad f \equiv 0; \quad u \in L_2(\Sigma). \qquad (3.3)$$

Then, the unique solution of (3.1) satisfies the interior regularity

$$w \in C([0, T]; H^\alpha(\Omega)), \qquad (3.4)$$
$$w_t \in C([0, T]; H^{\alpha-1}(\Omega)), \qquad (3.5)$$

and the boundary regularity

$$w|_\Sigma \in H^{2\alpha-1}(\Sigma). \qquad (3.6)$$

A fortiori, the map $u \to w|_\Sigma : L_2(\Sigma) \to L_2(\Sigma)$ is compact, since $\alpha > \frac{1}{2}$.

Theorem 3.2. ([L-T.5, Thm. A]) With reference to (3.1), assume

(a) (interior regularity) $w_0 = w_1 = 0; f \equiv 0$, and

$$u \in H^1(0, T; L_2(\Gamma)) \cap C([0, T]; H^{\alpha-\frac{1}{2}}(\Gamma)), \quad \text{and} \quad u(0) = 0. \qquad (3.7)$$

Then the unique solution of (3.1) satisfies

$$w \in C([0, T]; H^{\alpha+1}(\Omega)); \qquad (3.8)$$
$$w_t \in C([0, T]; H^\alpha(\Omega)). \qquad (3.9)$$

(b) (boundary regularity) Assume $w_0 = w_1 = 0$; $f \equiv 0$, and

$$u \in H^1(\Sigma) \quad \text{and} \quad u(0) = 0. \tag{3.10}$$

Then the unique solution of (3.1) satisfies

$$w|_\Sigma \in H^{2\alpha}(\Sigma). \tag{3.11}$$

Theorem 3.3.
(a) With reference to (3.1), assume

$$w_0 \in H^1(\Omega); \quad w_1 \in L_2(\Omega); \quad f \in L_2(Q); \quad u \equiv 0. \tag{3.12}$$

Then, the unique solution of (3.1) satisfies the interior regularity

$$w \in C([0, T]; H^1(\Omega)), \tag{3.13}$$
$$w_t \in C([0, T]; L_2(\Omega)), \tag{3.14}$$

and the boundary regularity ([L-T.3], [L-T.5, Thms. B and C]),

$$w|_\Sigma \in H^\beta(\Sigma). \tag{3.15}$$

(b) With A the $L_2(\Omega)$-realization of $-A(x, \partial)$ with homogeneous Neumann boundary
conditions

$$Ah = A(x, \partial)h; \quad \mathcal{D}(A) = \left\{ h \in H^2(\Omega) : \frac{\partial h}{\partial \nu_A} = 0 \right\}, \tag{3.16}$$

assume

$$w_0 \in \mathcal{D}(A); \quad w_1 \in H^1(\Omega); \quad f \in H^1(Q); \quad u \equiv 0. \tag{3.17}$$

Then, the unique solution of (3.1) satisfies the interior regularity

$$w \in C([0, T]; \mathcal{D}(A)), \tag{3.18}$$
$$w_t \in C([0, T]; H^1(\Omega)), \tag{3.19}$$

and the boundary regularity ([L-T.5, Thms. B and C]),

$$w|_\Sigma \in H^{\beta+1}(\Sigma). \tag{3.20}$$

Remarks 3.1.
(i)   In Theorem 3.1 the boundary regularity (3.6) does NOT follow from the interior
regularity (3.4), (3.5): it is an independent regularity result. The same holds true
in Theorem 3.2, under the assumption (3.10), which is stronger than assumption
(3.7) (as in Remark 2.1(ii), since $\alpha - \frac{1}{2} < \frac{1}{2}$ by (3.2)): the boundary regularity
(3.11) is an independent result, which does not follow from the interior regularity
(3.8), (3.9). Finally, similar comments apply to Theorem 3.3.
(ii) With reference to Theorem 3.2, we note explicitly that if one assumes

$$u \in H^{2\alpha-1,1}(\Sigma) = L_2(0, T; H^{2\alpha-1}(\Gamma)) \cap H^1(0, T; L_2(\Gamma)),$$

then [L-M.1; Thm. 3.1, $m = 1$, $j = 0$, p.19] u satisfies *a fortiori* the regularity condition in (3.7) (not the C.R. $u(0) = 0$), i.e.,

$$u \in C([0, T]; H^{\alpha - \frac{1}{2}}(\Gamma)).$$

(iii) In Theorem 3.3, the interior regularity (3.13), (3.14) [respect. (3.18), (3.19)] is, in fact, a standard result under the weaker assumption that $f \in L_1(0, T; L_2(\Omega))$ [respect. $f \in L_1(0, T; H^1(\Omega))$]. It is at the level of the boundary regularity (3.15) [respect. (3.20)] that the full strength of the assumption (3.12) [respect. (3.17)] on f is used, see [L-T.3].

(iv) By way of comparing, the best regularity results of the literature prior to [L-T.3] are those of [L-M.1], which state, in the case of assumptions (3.3) in Theorem 3.1, that, in fact, $\{w, w_t\} \in L_2(0, T; H^{\frac{1}{2}}(\Omega) \times H^{-\frac{1}{2}}(\Omega))$. This result does NOT allow one to conclude that $w|_\Sigma \in L_2(\Sigma)$, in contrast with (3.6). Thus, Theorem 3.1 represents an improvement of $\alpha - \frac{1}{2}$ in the space regularity of the solution, where $\alpha - \frac{1}{2}$ is at least equal to $\frac{1}{10} - \varepsilon$ in the general case (3.2a).

(In May 1984, during an exchange of correspondence, two proofs were given that $u \in L_2(\Sigma) \to w|_\Sigma \in L_2(\Sigma)$, one by J. L. Lions using Laplace transform, and one by the authors using Theorem 3.7 below, its dual version and interpolation, see [L-T.3, Remark 7.1]).

Similarly, the standard interior regularity (3.12) $\Rightarrow$ (3.13) in Theorem 3.3a would only imply by trace theory the boundary regularity $w|_\Sigma \in C([0, T]; H^{\frac{1}{2}}(\Gamma))$, whereby the boundary regularity (3.15) represents then an improvement of $\beta - \frac{1}{2}$ (equal at least to $\frac{1}{10}$ in the general case (3.2a)) in the space regularity. More on this in the *Notes* at the end of this section.

(v) Let now dim $\Omega = 1$. It is a standard elementary result that then

$$u \in L_2(\Sigma) \Rightarrow \begin{cases} \{w, w_t\} \in C([O, T] : H^1(\Omega) \times L_2(\Omega)) : & (3.21a) \\ w|_\Sigma \in H^1[0, T]. & (3.21b) \end{cases}$$

The above regularity result is false for dim $\Omega \geq 2$, see [Las.3], [L-T.3], [L-T.6] for a counterexample. Moreover, one can show by considering the canonical Laplacian $-A(x, \partial) = \Delta$ on the half-space, and hence on a parallelopiped, that in this case, if $w_0 = w_1 = 0$, $f \equiv 0$, then [L-T.6]

$$u \in L_2(\Sigma) \Rightarrow w \in H^{\frac{3}{4}}(Q), \quad \text{yet} \quad w \notin H^{\frac{3}{4} + \varepsilon}(Q), \qquad \forall \varepsilon > 0. \tag{3.22}$$

(vi) It was shown in [Sy.1] by means of geometric optics techniques and reproved in [L-T.6] by a direct Laplace-Fourier analysis, that in the canonical case of the Laplacian $-A(x, \partial) = \Delta$ on the half-space, with $u \equiv 0$, then

$$w_0 \in H^1(Q), \quad w_1 \in L_2(\Omega), \quad f \in L_2(Q) \Rightarrow w|_\Sigma \in H^1(\Sigma), \tag{3.23}$$

provided, in addition, the data $\{w_0, w_1, f\}$ are compactly supported away from the boundary Γ.

(vii) In the case of $A(x, \partial) = -\Delta$ on special geometries, where Ω is either a sphere or a parallelepiped, the above regularity results of Theorem 3.1 with $\alpha = \frac{2}{3}$, or $\alpha = \frac{3}{4} - \varepsilon$, are proved in [L-T.1] by direct eigenfunction expansion followed by Fourier transform. Later, [L-T.6] removed the ε by the technique in (v) via half-space analysis.

(viii) In the special case where the coefficients a_{ij} either do not depend on the normal variable or else do not depend on the tangential variable near Γ, we have the following results [Las.3], [L-T.3]:

$$w_0 = w_1 = 0; \quad f = 0; \quad u \in L_2(\Sigma) \Rightarrow \begin{cases} \{w, w_t\} \in C([0, T]; H^{\frac{2}{3}}(\Omega) \times H^{-\frac{1}{3}}(\Omega)) \\ \text{and} \\ w|_\Sigma \in H^{\frac{1}{3}}(\Sigma) \end{cases}$$

$$w_0 = w_1 = 0; \quad f \in L_2(\Sigma); \quad u = 0 \Rightarrow w|_\Sigma \in H^{\frac{2}{3}}(\Sigma),$$

i.e., we obtain the results of Theorem 3.1 and, respectively, of Theorem 3.3 with $\alpha = \beta = \frac{2}{3}$; i.e., the same regularity which was obtained in the case of $A(x, \partial) = -\Delta$ on a sphere in [L-T.1].

(ix) In the pseudo-differential proofs in [L-T.3] of the two preliminary results, Theorem 3.1 with $u \in L_2(\Sigma)$, and Theorem 3.3 with $f \in L_2(Q)$, it is crucially used that the coefficients a_{ij} are time independent. The subsequent functional analytic approach of [L-T.5] also refers to the time independent case.

Of the several possible interpolation results, we shall quote only one to be used in Section 6 (in the proof of Lemma 6.3). As noted in Remark 3.1(iii) we have *a fortiori* via Theorem 3.2:

$$\begin{matrix} u \in H^{2\alpha-1,1}(\Sigma) \\ u(0) = 0 \end{matrix} \Rightarrow \{w, w_t\} \in C([0, T]; H^\alpha(\Omega) \times H^{\alpha-1}(\Omega)). \qquad (3.24)$$

Interpolating between (3.24) and (3.3) $\Rightarrow$ (3.4), (3.5) of Theorem 3.1 with $0 < \theta \le \frac{1}{2}$ so that the C.R. $u(0) = 0$ is irrelevant, we obtain part (a) of the following corollary. Its part (b) is instead obtained by interpolating between (3.3) $\Rightarrow$ (3.6) of Theorem 3.1 and (3.10) $\Rightarrow$ (3.11) of Theorem 3.2.

Corollary 3.4. [L-T.5, Remark 3.6 and Corollary 4.3]
(a) With reference to (3.1), assume

$$w_0 = w_1 = 0; \quad f = 0; \quad u \in H^{(2\alpha-1)\theta,\theta}(\Sigma), \quad 0 \le \theta \le \frac{1}{2}. \qquad (3.25)$$

Then, the unique solution of (3.1) satisfies the interior regularity

$$\{w, w_t\} \in C([0, T]; H^{\alpha+\theta}(\Omega) \times H^{\alpha+\theta-1}(\Omega)). \qquad (3.26)$$

In particular, for $\theta = 1 - \alpha < \frac{1}{2}$ (by (3.2)), we have *a fortiori* with $w_0 = w_1 = 0$, $f = 0$:

$$u \in H^{1-\alpha}(\Sigma) \Rightarrow \{w, w_t\} \in C([0, T]; H^1(\Omega) \times L_2(\Omega)). \qquad (3.27)$$

(b) If $w_0 = w_1 = 0, f = 0$, then

$$u \in H^\theta(\Sigma) \Rightarrow w|_\Sigma \in H^{2\alpha-1+\theta}(\Sigma), \qquad 0 \le \theta \le \frac{1}{2}. \qquad (3.28)$$

We now consider the case of data $\{w_0, w_1, f\}$ "less regular than L_2".

Theorem 3.5. ([L-T.5, Thms. E and F]). With reference to (3.1) with $u \equiv 0$, assume

$$w_0 \in L_2(\Omega); \quad w_1 \in [H^1(\Omega)]'; \quad f \in [H^1(Q)]'. \qquad (3.29)$$

Then, the unique solution of (3.1) satisfies (where $1 - \alpha < \frac{1}{2}$)

$$w|_\Sigma \in H^{\alpha-1}(\Sigma) = [H^{1-\alpha}(\Sigma)]'. \tag{3.30}$$

Corollary 3.6. With reference to (3.1) with $u \equiv 0$, we have for all $0 \le \theta \le 1$,

$$\{w_0, w_1, f\} \in H^\theta(\Omega) \times [H^{1-\theta}(\Omega)]' \times [H^{1-\theta}(Q)]' \to H^{\alpha-1+\theta+(\beta-\alpha)\theta}(\Sigma). \qquad \Box \tag{3.31}$$

The proof of Corollary 3.6 follows by interpolating Theorem 3.3, Eq. (3.15), and Theorem 3.5.

Notes. *We have already noted in Remark 3.1(iv) that, say, with $w_0 = w_1 = 0$, $f = 0$ and $u \in L_2(\Sigma)$, the best regularity results available in the literature prior to [L-T.3] are those of [L-M.1] which assert that then $\{w, w_t\} \in L_2(0, T; H^{\frac{1}{2}}(\Omega) \times H^{-\frac{1}{2}}(\Omega))$ [L-M.1, vol. II, p. 120], or $\{w, w_t\} \in C([0, T]; H^{\frac{1}{2}-\varepsilon}(\Omega) \times [H^{\frac{1}{2}+\varepsilon}(\Omega)]')$ [L-M.1, vol. I, p. 120]; i.e., a 'gain' of $\frac{1}{2}$ in space regularity for w, versus the gain of $\alpha > \frac{1}{2}$ (in space regularity for w) asserted by Theorem 3.1. The work of [M.1] (which considers also the time-dependent case through a pseudo-differential approach in the style of [Sa.1], [Sa.2]) shares with [L-M.1] the same philosophy of an improvement of $\frac{1}{2}$ in Sobolev (space) regularity from u to w. The main result of [M.1] has, however, an assumption which is not symmetric in time and space. In the sequel we shall also make use of this result.*

Theorem 3.7. [M.1] With reference to (3.1), assume

$$w_0 = w_1 = 0; \quad f = 0; \quad u \in L_2(0, T; H^{\frac{1}{2}}(\Gamma)). \tag{3.32}$$

Then the unique solution of (3.1) satisfies

$$\{w, w_t\} \in C([0, T]; H^1(\Omega) \times L_2(\Omega)). \tag{3.33}$$

Remark 3.2. Theorem 3.7 should be compared with Corollary 3.4, Eq. (3.27). They both obtain the same conclusion by assuming, in the first case, $\frac{1}{2}$ regularity in space for u, and in the second case $(1 - \alpha)$ regularity in time for u, where $(1 - \alpha) \le \frac{2}{5}$ by (3.2). Actually, a stronger result than Eq. (3.27) is available, see [L-T.5, Eq. (1.22); or Corollary 3.4(i)]. $\Box$

To emphasize the '$\frac{1}{2}$-gain' regularity (for w in the interior) before [L-T.3] versus the role of the 'α-gain' regularity of the sharp theory to be exploited in subsequent sections, we shall interpolate between (3.32) $\Rightarrow$ (3.33) of Theorem 3.7 and the statement $u \in L_2(\Sigma) \Rightarrow \{w, w_t\} \in C([0, T]; H^{\frac{1}{2}}(\Omega) \times H^{-\frac{1}{2}}(\Omega))$ (weaker than (3.8), (3.9), i.e., the statement of [L-M.1] improved in time, which can be done via, e.g., an operator technique as in [L-T.7]). We obtain

Corollary 3.8. With reference to (3.1) assume

$$w_0 = w_1 = 0; \quad f = 0; \quad u \in L_2(0, T; H^{\theta/2}(\Gamma)), \quad 0 \le \theta \le 1. \tag{3.34}$$

Then the unique solution of (3.1) satisfies

$$\{w, w_t\} \in C([0, T]; H^{\frac{1+\theta}{2}}(\Omega) \times H^{\frac{\theta-1}{2}}(\Omega)). \tag{3.35}$$

Remark 3.3. In Part II of this paper, which deals with applications of the regularity theory, we shall repeatedly see that the theory of [L-M.1], [M.1] with a gain of $\alpha = \frac{1}{2}$

is inadequate to show a number of results. In contrast, it will be *critical* to have that in fact $\alpha > \frac{1}{2}$, the particular value of α being of less importance. An example of the critical importance of $\alpha > \frac{1}{2}$ versus $\alpha = \frac{1}{2}$, was already seen below (3.6) in asserting that the map $w_0 = w_1 = 0; \dot{f} = 0; u \in L_2(\Sigma) \to w|_\Sigma \in L_2(\Sigma)$ is compact.

4. Cosine/Sine (Semigroup) Representation Formulae of the Solutions

The goal of the present section is two-fold:

(i) we first provide representation formulae for the solutions of the mixed hyperbolic problems (2.1) (Dirichlet) and (3.1) (Neumann), which are based on cosine/sine (semigroup) theory;

(ii) we next give the preceding regularity theory of Sections 2 and 3 an operator interpretation or reformulation, which we find particularly convenient as will be seen in subsequent sections.

4.1. Dirichlet Case

Preliminaries. Throughout this subsection, we let A be the (closed) operator $L_2(\Omega) \supset \mathcal{D}(A) \to L_2(\Omega)$ defined by

$$Ah = A(x, \partial)h, \qquad \mathcal{D}(A) = H^2(\Omega) \cap H_0^1(\Omega), \qquad (4.1)$$

which corresponds to problem (2.1) with homogeneous boundary conditions. Then, $-A$ generates a s.c. cosine operator $C(t)$ with $S(t) = \int_0^t C(\tau)d\tau$ on $L_2(\Omega)$, $t \in R$, [Fa.2]. The fractional powers A^θ, $0 < \theta < 1$, of A are well defined, if we assume, without loss of generality, that after an innocous translation the spectrum of A is all in (a suitable parabolic sector with) $\text{Re } \lambda > 0$. We shall abusively call $S(t)$ the corresponding sine operator (although, technically, it is $A^{\frac{1}{2}} S(t)$ that has the right credentials for being called sine operator). For $s \geq 0$ we make $\mathcal{D}(A)$ a Hilbert space with norm (equivalent to the graph norm)

$$|z|_{\mathcal{D}(A^s)} = |A^s z|_{L_2(\Omega)}, \qquad s \geq 0. \qquad (4.2)$$

Moreover, still for $s > 0$, if A^* is the $L_2(\Omega)$-adjoint of A, we denote by $[\mathcal{D}(A^s)]'$, respect. $[\mathcal{D}(A^{*s})]'$, the dual space of $\mathcal{D}(A^s)$, respect. $\mathcal{D}(A^{*s})$, with respect to the $L_2(\Omega)$-topology, which are Hilbert spaces when endowed with norm

$$|z|_{[\mathcal{D}(A^s)]'} = |A^{*-s}z|_{L_2(\Omega)}; \qquad |z|_{[\mathcal{D}(A^{*s})]'} = |A^{-s}z|_{L_2(\Omega)}. \qquad (4.3)$$

Though the original $A(x, \partial)$ in (1.4) has symmetric coefficients we shall keep both notations A and A^* to cover simultaneously the case where $A(x, \partial)$ is perturbed by a first-order operator with smooth coefficients.) The following identification (with equivalent norms) is known [G.1], [F.2], [Las.1; Appendix],

$$\mathcal{D}(A^\theta) = H_0^{2\theta}(\Omega), \quad \text{hence} \quad [\mathcal{D}(A^\theta)]' = H^{-2\theta}(\Omega), \quad \theta \neq 1/4, \quad 0 < \theta < 3/4. \quad (4.4)$$

The original operator $-A$ in (4.1) can then be restricted (respect. extended) to act on the space $\mathcal{D}(A^s)$, $s > 0$, (respect. on the space $[\mathcal{D}(A^s)]'$, $s > 0$, by isomorphism techniques) to preserve its property of being here the generator of a s.c. restricted (respect. extended) cosine operator. For sake of simplicity of notation, the restriction (respect. extension) of the original operators A, $C(t)$, $S(t)$ will be indicated by the same symbols. Using (4.4), one then obtains from known results on $L_2(\Omega)$ ($\theta = 0$):

$$C(t) : \text{continuous } H^{-\theta}(\Omega) \to C([0, T]; H^{-\theta}(\Omega)), \quad 0 \le \theta \le 3/2, \ \theta \ne 1/2; \quad (4.5)$$
$$S(t) : \text{continuous } H^{-\theta}(\Omega) \to C([0, T]; H_0^{1-\theta}(\Omega)), \quad 0 \le \theta \le 3/2, \ \theta \ne 1/2. \quad (4.6)$$

Where $H^{-\theta}(\Omega) = [\mathcal{D}(A^{\theta/2})]'$, $H_0^{1-\theta}(\Omega) = \mathcal{D}(A^{\frac{1-\theta}{2}})$, $\theta \ne \frac{1}{2}$; the case $\theta = \frac{1}{2}$ is

$$C(t) : \text{continuous } [\mathcal{D}(A^{\frac{1}{4}})]' \to C([0, T]; [\mathcal{D}(A^{\frac{1}{4}})]'); \quad (4.7)$$
$$S(t) : \text{continuous } [\mathcal{D}(A^{\frac{1}{4}})]' \to C([0, T]; \mathcal{D}(A^{\frac{1}{4}})). \quad (4.8)$$

We next introduce the Dirichlet map D defined by

$$h = Dv \iff \{-A(x, \partial)h = 0 \text{ in } \Omega; \ h|_\Gamma = v \text{ on } \Gamma\}, \quad (4.9)$$

where under the present assumption we have without loss of generality for $A(x, \partial)$ that the elliptic problem in (4.9) has a unique solution. The following elliptic regularity result is known [L-M.1, p. 189], [N.1], [K.1],

$$D : \text{continuous } H^s(\Gamma) \to H^{s+\frac{1}{2}}(\Omega), \quad s \in R, \quad (4.10)$$
$$: \text{continuous } L_2(\Gamma) \to H^{\frac{1}{2}}(\Omega) \subset H^{\frac{1}{2}-2\rho}(\Omega) = \mathcal{D}(A^{\frac{1}{4}-\rho}), \quad \rho > 0. \quad (4.11)$$

By using Green's second theorem, one obtains [L-L-T.1, p. 171, p. 185]

$$D^* A^* h = \frac{\partial h}{\partial v_{A^*}} = \sum_{i,j}^n a_{ij} v_i \frac{\partial}{\partial x_j}, \quad h \in \mathcal{D}(A^*) = \mathcal{D}(A), \quad (4.12)$$

where $(Du, z)_{L_2(\Omega)} = (u, D^* z)_{L_2(\Gamma)}$, and where we shall always write $D^* A^*$ to mean, if necessary, its extensions $(AD)^*$.

Remark 4.1. With the symmetricity condition $a_{ij} = a_{ji}$ on the coefficients, as assumed in (1.4), we have that the co-normal derivative is

$$\frac{\partial}{\partial v_{A^*}} = \frac{\partial}{\partial v_A} = \sum_{i,j}^n a_{ij} v_i \frac{\partial}{\partial x_j}. \quad (4.13)$$

If now $\psi = 0$ on Γ, then

$$\frac{\partial \psi}{\partial x_j} = v_j \frac{\partial \psi}{\partial v}, \quad \psi|_\Gamma = 0, \quad (4.14)$$

while in the case of the solution w of problem (2.1), where $w|_\Gamma = u$ we have [L-L-T.1, p. 160-1]

$$\frac{\partial w}{\partial x_j} = v_j \frac{\partial w}{\partial v} + T_j u \quad w|_\Gamma = 0, \quad (4.15)$$

$$T_j = \text{first-order differential operator on } \Gamma$$
$$\text{with smooth coefficients (for } \Gamma \text{ smooth)}.$$

Thus, (4.13) and (4.15) give

$$D^*A^*w = \frac{\partial w}{\partial \nu_{A^*}} = \frac{\partial w}{\partial \nu_A} = \sum_{i,j}^n a_{ij}\nu_i\nu_j\frac{\partial w}{\partial \nu} + \sum_{i,j}^n a_{ij}\nu_i T_j u, \quad w|_\Gamma = u; \tag{4.16}$$

$$|D^*A^*w|^2 = \left|\frac{\partial w}{\partial \nu_A}\right|^2 \geq (c - \varepsilon)\left|\frac{\partial w}{\partial \nu}\right|^2 + \mathcal{O}(|T_j u|^2), \quad w|_\Gamma = u; \tag{4.17}$$

after multiplying (4.16) by $\frac{\partial w}{\partial \nu}$ and recalling the ellipticity condition (1.4b).

Abstract models of problem (2.1). Second-order model. By the definition (4.5) of D, the original problem (2.1a), (2.1c) can be rewritten as

$$w_{tt} = -A(x, \partial)(w - Du) + f \quad \text{in } Q, \tag{4.18a}$$
$$[w - Du]_\Sigma \equiv 0 \qquad \text{in } \Sigma, \tag{4.18b}$$

i.e., by (4.1) abstractly as

$$w_{tt} = -A(w - Du) + f \quad \text{on } L_2(\Omega); \qquad \text{or} \tag{4.19a}$$
$$w_{tt} = -Aw + ADu + f \quad \text{on, say,} \quad [\mathcal{D}(A^{*\frac{1}{4}+\rho})]' \tag{4.19b}$$

where in passing from the factor form in (4.19a) to the additive form in (4.19b) we are actually using the isomorphic extension, say, $A : L_2(\Omega) \to [\mathcal{D}(A^*)]'$. The integral version, which corresponds to the differential versions in (4.19), is

$$w(t) = w(t; w_0, w_1) = C(t)w_0 + S(t)w_1 + (Lu)(t) + (Kf)(t); \tag{4.20}$$
$$w_t(t) = w_t(t; w_0, w_1) = -AS(t)w_0 + C(t)w_1 + \left(\frac{dLu}{dt}\right)(t) + \left(\frac{dKf}{dt}\right)(t); \tag{4.21}$$
$$(Lu)(t) = A\int_0^t S(t - \tau)Du(\tau)d\tau; \quad \left(\frac{dLu}{dt}\right)(t) = A\int_0^t C(t - \tau)Du(\tau)d\tau; \tag{4.22}$$
$$(Kf)(t) = \int_0^t S(t - \tau)f(\tau)d\tau; \quad \left(\frac{dKf}{dt}\right)(t) = \int_0^t C(t - \tau)f(\tau)d\tau; \tag{4.23}$$

First-order model. The first-order differential model corresponding to the second-order models in (4.19) is

$$\frac{d}{dt}\begin{vmatrix} w \\ w_t \end{vmatrix} = \mathcal{A}\begin{vmatrix} w \\ w_t \end{vmatrix} + \mathcal{B}u, \quad \text{on, say,} \quad Y = Y_1 \times Y_2, \tag{4.24}$$

where, if we take $u \in L_2(0, T; L_2(\Gamma))$, the natural spaces are

$$Y_1 = L_2(\Omega); \quad Y_2 = H^{-1}(\Omega) = [\mathcal{D}(A^{\frac{1}{2}})]'; \tag{4.25}$$
$$\mathcal{A} = \begin{vmatrix} 0 & I \\ -A & O \end{vmatrix}; \quad \mathcal{D}(\mathcal{A}) = \mathcal{D}(A^{\frac{1}{2}}) \times L_2(\Omega); \tag{4.26}$$

$$\mathcal{B}u\begin{vmatrix} 0 \\ ADu \end{vmatrix}; \quad \mathcal{A}^{-1}\mathcal{B}u = \begin{vmatrix} -Du \\ 0 \end{vmatrix}; \tag{4.27}$$

and $\mathcal{B}$ may be viewed as an operator,

$$\mathcal{B} : \text{continuous } L_2(\Gamma) \rightarrow [\mathcal{D}(\mathcal{A}^*)]'; \quad \text{equivalently} \tag{4.28a}$$

$$\mathcal{A}^{-1}\mathcal{B} : \text{continuous } U = L_2(\Gamma) \rightarrow Y \quad (\text{in fact } \rightarrow \mathcal{D}(A^{\frac{1}{4}-\rho}) \times L_2(\Omega)). \tag{4.28b}$$

We have

$$e^{\mathcal{A}t} = \begin{vmatrix} C(t) & S(t) \\ -AS(t) & C(t) \end{vmatrix}; \tag{4.29}$$

$$(\mathcal{L}u)(t) = \int_0^t e^{\mathcal{A}(t-\tau)}\mathcal{B}u(\tau)d\tau \tag{4.30}$$

$$= \begin{vmatrix} (\mathcal{L}u)(t) \\ (\frac{d\mathcal{L}u}{dt})(t) \end{vmatrix} = \begin{vmatrix} A \int_0^t S(t-\tau)Du(\tau)d\tau \\ A \int_0^t C(t-\tau)Du(\tau)d\tau \end{vmatrix}; \tag{4.31}$$

$$\mathcal{B}^* \begin{vmatrix} z_1 \\ z_2 \end{vmatrix} = D^*A^*A^{-\frac{1}{2}}A^{*-\frac{1}{2}}z_2; \tag{4.32}$$

$$(\mathcal{B}u, z)_Y = (ADu, z_2)_{[\mathcal{D}(A^{\frac{1}{2}})]'} = (u, \mathcal{B}^*z)_{L_2(\Gamma)}, \quad \text{by (4.3)}; \tag{4.33}$$

$$\mathcal{B}^*e^{\mathcal{A}^*t} \begin{vmatrix} y_1 \\ y_2 \end{vmatrix} = D^*A^*S^*(t)y_1 + D^*C^*(t)y_2, \quad [y_1, y_2] \in Y. \tag{4.34}$$

Regularity results of Section 2 in operator form. By using (4.4) and its dual form, as well as (4.12), (4.17) and (4.20)–(4.23), it is possible to rewrite, or translate, the regularity results of Section 2 in terms of the operators introduced in this section. We only list below a selected sample of such results, and others can likewise be given [L-L-T.1, Sect. 3], [L-T.8, Sect. 3].

Theorem 4.1. (Boundary regularity due to w_0, w_1.) With respect to the operators introduced above, we have

$$D^*A^*S(t) : \text{continuous } L_2(\Omega) \rightarrow L_2(\Sigma); \tag{4.35}$$

$$D^*A^*C(t) : \text{continuous } H_0^1(\Omega) = \mathcal{D}(A^{\frac{1}{2}}) \rightarrow L_2(\Sigma); \tag{4.36}$$

$$D^*A^*S(t) : \text{continuous } H^{-1}(\Omega) = [\mathcal{D}(A^{\frac{1}{2}})]' \rightarrow H^{-1}(\Sigma); \tag{4.37}$$

$$D^*A^*C(t) : \text{continuous } L_2(\Omega) \rightarrow H^{-1}(\Sigma); \tag{4.38}$$

$$D^*A^*S(t) : \text{continuous } H_0^1(\Omega) = \mathcal{D}(A^{\frac{1}{2}}) \rightarrow H^1(\Sigma); \tag{4.39}$$

$$D^*A^*C(t) : \text{continuous } \mathcal{D}(A) \rightarrow H^1(\Sigma). \tag{4.40}$$

The same results hold true with $S(t)$, $C(t)$ replaced, respectively, by $S^*(t)$, $C^*(t)$.

In view of (4.34), statements (4.35)–(4.40) can be rewritten as

$$\mathcal{B}^*e^{\mathcal{A}^*t} : \text{continuous} \begin{cases} H_0^1(\Omega) \times L_2(\Omega) \rightarrow H^1(\Sigma) : & (4.41) \\ L_2(\Omega) \times H^{-1}(\Omega) \rightarrow L_2(\Sigma) : & (4.42) \\ H^{-1}(\Omega) \times [\mathcal{D}(A)]' \rightarrow H^{-1}(\Sigma) : & (4.43) \end{cases}$$

Proof. We return to (4.20) with $f \equiv 0$ and $u \equiv 0$: then (4.37), (4.38) are a restatement of (2.8) in Theorem 2.1 via (4.16), (4.17); similarly, (4.35), (4.36) are a restatement of

(2.16) in Theorem 2.2 (note that the c.c. (2.12) gives $w_0|_\Gamma = 0$); finally, (4.39), (4.40) are a restatement of (2.24) in Theorem 2.3 (note that the c.c. (2.20) gives $w_0|_\Gamma = w_1|_\Gamma = 0$).

Theorem 4.2. (Interior and boundary regularity due to u.) With reference to the operators in (4.22), (4.24), and (4.16), we have

$$L : \text{continuous } L_2(\Sigma) \to C([0, T]; L_2(\Omega)); \tag{4.44}$$

$$\frac{dL}{dt} : \text{continuous } L_2(\Sigma) \to C([0, T]; H^{-1}(\Omega)); \tag{4.45}$$

$$D^*A^*L : \text{continuous} \begin{cases} L_2(\Sigma) \to H^{-1}(\Sigma) : & (4.46) \\[2mm] \left.\begin{cases} H^1(\Sigma) \\ +c.c. \; u|_{t=0} = 0 \end{cases}\right\} \to L_2(\Sigma) : & (4.47) \\[2mm] \left.\begin{cases} H^2(\Sigma) \\ +c.c \; u|_{t=0} = u_t|_{t=0} = 0 \end{cases}\right\} \to H^1(\Sigma). & (4.48) \end{cases}$$

Proof. Conclusions (4.44), (4.45) are a restatement of (2.6), (2.7), in Theorem 2.1 via (4.21); conclusions (4.46), (4.47), (4.48) are a restatement of (2.8) in Theorem 2.1; (2.16) in Theorem 2.2 with c.c. (2.12); and (2.24) in Theorem 2.3 with c.c. (2.20). $\quad\blacksquare$

Theorem 4.3. (Interior and boundary regularity due to f.) With reference to (4.23), (4.16), we have

$$K, \frac{dK}{dt} : \text{cont.} \begin{cases} L_1(0, T : H^{-1}(\Omega)) \to C([0, T] : L_2(\Omega) \times H^{-1}(\Omega)), & (4.49) \\ L_1(0, T : L_2(\Omega)) \to C([0, T] : H_0^1(\Omega) \times L_2(\Omega)), & (4.50) \\ L_1(0, T : H_0^1(\Omega)) \to C([0, T] : [H^2(\Omega) \cap H_0^1(\Omega)] \times H_0^1(\Omega)), & (4.51) \end{cases}$$

where we recall (4.1) and (4.4) with $\theta = \frac{1}{2}$,

$$D^*A^*K : \text{continuous} \begin{cases} L_1(0, T : H^{-1}(\Omega)) \to H^{-1}(\Sigma). & (4.52) \\ L_1(0, T : L_2(\Omega)) \to L_2(\Sigma), & (4.53) \\ L_1(0, T : H^1(\Omega)) \to H^1(\Sigma). & (4.54) \end{cases}$$

Proof. Conclusions (4.49), (4.50), (4.51) are a restatement of (2.6), (2.7) in Theorem 2.1; (2.13), (2.14) in Theorem (2.2); (2.21), (2.22) in Theorem 2.3; where $w_0 = w_1 = 0$; $u \equiv 0$. Similarly, conclusions (4.52), (4.53), (4.54) are a restatement of (2.8) in Theorem 2.1; (2.16) in Theorem 2.2; and (2.24) in Theorem 2.3. $\quad\blacksquare$

4.2. Neumann Case

Preliminaries. We recall from (3.16) that throughout this subsection A is defined by

$$Ah = A(x, \partial)h, \qquad \mathcal{D}(A) = \left\{ h \in H^2(\Omega) : \frac{\partial h}{\partial \nu_A} = 0 \right\}. \tag{4.55}$$

Then, $-A$ will likewise generate a s.c. cosine operator again denoted by $C(t)$ on $L_2(\Omega)$, $t \in R$. Modulo an innocous translation, we shall again assume without loss of generality that the spectrum of A is all contained in (a parabolic sector in) $\text{Re}\,\lambda > 0$. Then, the following identification (with equivalent norms) is known [G.1], [Fu.1], [Las.1, App.]

$$\mathcal{D}(A^\theta) = H^{2\theta}(\Omega), \qquad 0 \le \theta < \frac{3}{4}. \tag{4.56}$$

As in Section 4.1, we may now write

$$C(t) : \text{ continuous } \mathcal{D}(A^s) \to C([0, T]; \mathcal{D}(A^s)), \tag{4.57}$$

$$S(t) : \text{ continuous } \mathcal{D}(A^s) \to C([0, T]; \mathcal{D}(A^{s+\frac{1}{2}})), \tag{4.58}$$

where in (4.57), (4.58) we have used the conventional notation $\mathcal{D}(A^{-s}) = [\mathcal{D}(A^s)]'$ for $s > 0$.

We next introduce the Neumann map N. Let the null space $\mathcal{N}(A)$ of A in (4.55) be trivial: $\mathcal{N}(A) = \{0\}$, as assumed without loss of generality. Then, analogously to (4.9) we define

$$h = Nv \iff \left\{ -A(x, \partial)h = 0 \text{ in } \Omega; \ \frac{\partial h}{\partial \nu_A}\big|_\Gamma = v \text{ on } \Gamma \right\}. \tag{4.59}$$

In cases like $A(x, \partial) = -\Delta$ where then $\mathcal{N}(A)$ is the one-dimensional manifold of the constant functions, one may define the operator $\overline{N}$, in place of N:

$$h = \overline{N}v \iff \left\{ (-A(x, \partial) - \lambda_0)h = 0 \text{ in } \Omega; \ \frac{\partial h}{\partial \nu_A}\big|_\Gamma = v \text{ on } \Gamma \right\}$$

for a suitably large positive λ_0 as to have uniqueness of the corresponding elliptic problem. For the purposes of this section, we may assume $\mathcal{N}(A) = \{0\}$.] The counterpart of (4.10), (4.11) is the following elliptic regularity [L-M.1, p. 189], [N.1], [K.1]:

$$N : \text{ continuous } H^s(\Gamma) \to H^{s+\frac{3}{2}}(\Omega) \tag{4.60}$$

$$: \text{ continuous } L_2(\Gamma) \to H^{\frac{3}{2}}(\Omega) \subset H^{\frac{3}{2}-2\rho}(\Omega) = \mathcal{D}(A^{\frac{3}{4}-\rho}), \quad \rho > 0. \tag{4.61}$$

The counterpart of (4.12) is now

$$N^* A^* h = h|_\Gamma, \qquad h \in \mathcal{D}(A^*), \tag{4.62}$$

which is again obtained by Green's second theorem, where $(N^* u, z)_{L_2(\Omega)} = (u, Nz)_{L_2(\Gamma)}$.

Abstract models for problem (3. 1). Scond-order model. By the definition (4.59) of N, the original problem (3.1a), (3.1c) can be rewritten as

$$w_{tt} = -A(x, \partial)(w - Nu) + f \qquad \text{in } Q, \tag{4.63}$$

$$\frac{\partial}{\partial \nu_A}(w - Nu) = 0 \qquad \text{in } \Sigma, \tag{4.64}$$

i.e., by (4.55) abstractly as

$$w_{tt} = -A(w - Nu) + f \text{ on } L_2(\Omega); \qquad \text{or} \qquad w_{tt} = -Aw + ANu + f \text{ on } [\mathcal{D}(A^{*\frac{3}{4}+\rho})]', \tag{4.65}$$

where again on the right of (4.65) we are actually using the extension, say $A : L_2(\Omega) \rightarrow [\mathcal{D}(A^*)]'$. The integral version which corresponds to the differential versions in (4.65) is

$$w(t) = w(t; w_0, w_1) = C(t)w_0 + S(t)w_1 + (Lu)(t) + (Kf)(t); \tag{4.66}$$

$$w_t(t) = w_t(t; w_0, w_1) = -AS(t)w_0 + C(t)w_1 + \left(\frac{dLu}{dt}\right)(t) + \left(\frac{dKf}{dt}\right)(t); \tag{4.67}$$

$$(Lu)(t) = \int_0^t AS(t - \tau)Nu(\tau)d\tau; \qquad \left(\frac{dLu}{dt}\right)(t) = \int_0^t AC(t - \tau)Nu(\tau)d\tau; \tag{4.68}$$

$$(Kf)(t) = \int_0^t S(t - \tau)f(\tau)d\tau; \qquad \left(\frac{dKf}{dt}\right)(t) = \int_0^t C(t - \tau)f(\tau)d\tau; \tag{4.69}$$

First-order model. The first-order differential model corresponding to the second-order models in (4.65) is formally the same as the one in Section 4.1 for the Dirichlet case, except that now we use the operator A in (4.55). We rewrite the second-order abstract equation in (4.65) as first-order system,

$$\frac{d}{dt}\begin{vmatrix} w \\ w_t \end{vmatrix} = A\begin{vmatrix} w \\ w_t \end{vmatrix} + \mathcal{B}u, \tag{4.70}$$

where now, with a view toward applications in Sections 5, 6, 8, we take

$$Y = Y_1 \times Y_2; \quad Y_1 = H^1(\Omega); \quad Y_2 = L_2(\Omega); \tag{4.71}$$

$$\mathcal{A} = \begin{vmatrix} 0 & I \\ -A & 0 \end{vmatrix}, \qquad \mathcal{D}(\mathcal{A}) = \mathcal{D}(A) \times \mathcal{D}(A^{\frac{1}{2}}); \tag{4.72}$$

A the operator in (4.55), while $\mathcal{B}$ is now

$$\mathcal{B}u = \begin{vmatrix} 0 \\ ANu \end{vmatrix}; \qquad \mathcal{A}^{-1}\mathcal{B}u = \begin{vmatrix} -Nu \\ 0 \end{vmatrix}; \tag{4.73}$$

$$\mathcal{A}^{-1}\mathcal{B} : \text{ continuous } L_2(\Gamma) \rightarrow \mathcal{D}(A^{\frac{3}{4}-\rho}) \times L_2(\Omega). \tag{4.74}$$

We have

$$e^{\mathcal{A}t} = \begin{vmatrix} C(t) & S(t) \\ -AS(t) & C(t) \end{vmatrix}, \tag{4.75}$$

where now $C(t)$ is generated by A in (4.55), and

$$(\mathcal{L}u)(t) = \int_0^t e^{\mathcal{A}(t-\tau)}\mathcal{B}u(\tau)d\tau \tag{4.76}$$

$$= \begin{vmatrix} (Lu)(t) \\ \left(\frac{dLu}{dt}\right)(t) \end{vmatrix} = \begin{vmatrix} \int_0^t AS(t - \tau)Nu(\tau)d\tau \\ \int_0^t AC(t - \tau)Nu(\tau)d\tau \end{vmatrix}; \tag{4.77}$$

$$\mathcal{B}^*\begin{vmatrix} z_1 \\ z_2 \end{vmatrix} = N^*A^*z_2 = z_2|_\Gamma; \tag{4.78}$$

$$(\mathcal{B}u, z)_y = (ANu, z_2)_{L_2(\Omega)} = (u, N^*A^*z_2)_{L_2(\Gamma)} = (u, \mathcal{B}^*z_2)_{L_2(\Gamma)}; \tag{4.79}$$

$$\mathcal{B}^* e^{A^* t} \begin{vmatrix} y_1 \\ y_2 \end{vmatrix} = -N^* A^* A^* S^*(t) y_1 + N^* A^* C^*(t) y_2. \tag{4.80}$$

Regularity results of Section 3 in operator form. We now reformulate in operator form some selected regularity results of Section 3, using the framework introduced above.

Theorem 4.4. (Boundary regularity due to w_0, w_1.) With reference to (4.64), we have

$$N^* A^* S(t) : \text{continuous } L_2(\Omega) \to H^\beta(\Sigma); \tag{4.81}$$
$$N^* A^* C(t) : \text{continuous } H^1(\Omega) = \mathcal{D}(A^{\frac{1}{2}}) \to H^\beta(\Sigma); \tag{4.82}$$
$$N^* A^* S(t) : \text{continuous } H^1(\Omega) = \mathcal{D}(A^{\frac{1}{2}}) \to H^{\beta+1}(\Sigma); \tag{4.83}$$
$$N^* A^* C(t) : \text{continuous } \mathcal{D}(A) \to H^{\beta+1}(\Sigma); \tag{4.84}$$
$$N^* A^* S(t) : \text{continuous } [H^1(\Omega)]' = [\mathcal{D}(A^{\frac{1}{2}})]' \to H^{\alpha-1}(\Sigma); \tag{4.85}$$
$$N^* A^* C(t) : \text{continuous } L_2(\Omega) \to H^{\alpha-1}(\Sigma). \tag{4.86}$$

The same regularity results hold true with $C(t)$, $S(t)$ replaced, respectively, by $C^*(t)$, $S^*(t)$. Also, in view of (4.80) the above results can be rewritten as

$$\mathcal{B}^* e^{A^* t} : \text{continuous} \begin{cases} \mathcal{D}(A) \times \mathcal{D}(A^{\frac{1}{2}}) \to H^\beta(\Sigma) : & \tag{4.87} \\ \mathcal{D}(A^{\frac{3}{2}}) \times \mathcal{D}(A) \to H^{\beta+1}(\Sigma) : & \tag{4.88} \\ \mathcal{D}(A^{\frac{1}{2}}) \times L_2(\Omega) \to H^{\alpha-1}(\Sigma). & \tag{4.89} \end{cases}$$

Proof. Conclusions (4.81), (4.82) are a reformulation of (3.15) in Theorem 3.3 via (4.66) with $f \equiv 0$, $u \equiv 0$; similarly, conclusions (4.83), (4.84) are a reformulation of (3.17) in Theorem 3.3; finally, conclusions (4.85), (4.86) are a reformulation of (3.26) in Theorem 3.4. □

Theorem 4.5. (Interior and boundary regularity due to u.) With reference to (4.62), (4.68), we have

$$L : \text{continuous } L_2(\Sigma) \to C([0, T]; H^\alpha(\Omega)); \tag{4.90}$$
$$\frac{dL}{dt} : \text{continuous } L_2(\Sigma), \quad C([0, T]; H^{\alpha-1}(\Omega)); \tag{4.91}$$
$$L|_\Sigma = N^* A^* L : \text{continuous } L_2(\Sigma) \to H^{2\alpha-1}(\Sigma). \tag{4.92}$$

Proof. . This is a reformulation of Theorem 3.1. □

Remark 4.2. The following additional regularity results should be noted [L-T.2], [L-T.5]:

$$\left. \begin{array}{l} N^* A^{*1+\alpha/2} S^*(t) \\ N^* A^{*\frac{1}{2}+\alpha/2} C^*(t) \end{array} \right\} : \text{continuous } L_2(\Omega) \to L_2(\Sigma). \quad \square \tag{4.93}$$

Notes. *The abstract operator formulation for second-order equations of this section was introduced by the authors in [Tr.1], [L-T.1] and used successfully since in a series of papers dealing with a variety of different topics, including regularity (see e.g., [L-T.1], [L-T.2] for the Dirichlet case and [L-T.1], [L-T.5] for the Neumann case). Applicability of these abstract models includes also, 'mutatis mutandis', the case of plate-like equations (Petrovski type), where the differential operator is of order four and two boundary conditions are available. The corresponding model for first-order hyperbolic systems* *is contained in* [C-L.1]. *These explicit abstract models, whether in differential or in integral form via cosine/sine theory* [Fa.2] *(hence semigroup theory), have proven to be very useful, not only in handling specific problems, but also in providing a unified abstract framework approach to mixed problems for partial differential equations. Illustrations of both features are contained in subsequent sections, which show the usefulness of this approach as well. The integral form versions* (4.30), (4.31) *or* (4.76), (4.77), *are of the 'variation of constant' type. They were later arrived at, independently, in* [Am.1]. *The introduction of the models of this section in both differential and integral form benefited from the original contribution of* [Bal. 1] *(which pushed further an idea of* [Fa.1] *for both first- and second-order problems), where the abstract model in integral form* $A \int_0^t \exp[A(t - \tau)]Du(\tau)d\tau$ *is introduced for a parabolic problem with u in the Dirichlet B.C. and where the operator* $A = \Delta$ *with zero B.C.) generates an analytic semigroup* $\exp(At)$. *It was then proved in* [Bal.1] *in the special case of* Ω *being a square (via eigenfunctions expansion) and subsequently in* [W.1] *in the general analytic semigroup case by using the theory of intermediate spaces as in* [B-B.1] *that the bound* (#) : $|Ae^{At}D| = \mathcal{O}(t^{-\frac{3}{4}-\varepsilon})$ *holds* $0 < t \le 1$, $\forall \varepsilon > 0$, *in the* $L_2(\Omega)$-*uniform norm. That this important bound* (#) *can be directly and more easily proved by using* (4.11) *was observed in* [Tr.2], [Tr.3], [Tr.4]. *Note that* (4.11) *is based on elliptic regularity plus identification of domains of appropriate fractional powers with Sobolev spaces. In addition to re-proving* (#), *this introduction of fractional powers was proven to be a useful technical device in the analysis.*

We conclude by pointing out a drastic difference in the role played by abstract models in hyperbolic/plate mixed problems (second order in time) on the one hand, and parabolic mixed problems (first order in time) on the other.

In the latter case, the combination of semigroup methods with elliptic theory and statements like (4.11) *is sufficient to provide (or re-prove) optimal regularity results for parabolic mixed problems, see* [Las.1] *and say* [L-T.9]–[L-T.10]. *This theory includes the 'Hilbert theory' of, say,* [L-M.1], *which is obtained by different means.*

Not so, however, for hyperbolic/plate mixed problems. Here, the first crucial step or building block of a regularity theory comes from purely partial differential equation methods *(energy methods, either in calculus form as in the case of the Dirichlet problem of Section 2, see* Notes *of Section 2; or else in pseudo-differential form for the Neumann problem of Section 3, see* Notes *in Section 3). In this case of hyperbolic/plate mixed problems, abstract operator methods provide useful tools only at a* subsequent *level (for higher/lower data, duality or transposition, etc.), after a key preliminary regularity result has been obtained from energy methods, see e.g.,* [L-L-T.1] *in the Dirichlet case and* [L-T.5] *in the Neumann case.*

Part II: Applications

Part II is devoted to the application of the preceding regularity theory to booth some non-linear p.d.e. problems and to some control problems. While there is no space, of course, for inclusion of the proofs of all results, we shall selectively put our emphasis in pointing out the key role played by the regularity theory; and in particular, in the Neumann case, in singling out the critical role of the sharp regularity theory with $\alpha > \frac{1}{2}$ over the earlier theory with $\alpha = \frac{1}{2}$ in the successful solution of certain problems. References for further details are also provided. Another goal of these sections is to illustrate the usefulness of the abstract operator models, introduced in Section 4, in the technical analysis.

5. Well-posedness of Semi-linear Wave Equations with Neumann Boundary Conditions

5.1. Statement of Local Well-posedness

Throughout this section, we let Ω be a smooth bounded domain in R^n, $n \leq 3$. Moreover, let g be a scalar function of class C^3. We consider the semi-linear mixed problem

$$\begin{cases} w_{tt} - \Delta w + w = 0 & \text{in } (0, T] \times \Omega)Q : & (5.1a) \\ w(0, \cdot) = w_0 : w_t(o, \cdot) = w_1 & \text{in } \Omega : & (5.1b) \\ \frac{\partial w}{\partial \nu}|_\Sigma = g(w|_\Sigma) & \text{in } (0, T] \times \Gamma = \Sigma. & (5.1c) \end{cases}$$

Following [L-S.1], our goal is to provide local (in time) existence results for the solutions to problem (5.1) for a large class of nonlinear functions g. In doing so, we shall employ the functional analytic models described in Section 4, as well as the regularity theory of the corresponding linear problem of Section 3. Three (local) well-posedness results will be presented: Theorem 5.1 which provides well-posedness for (smoother) solutions $\{w, w_t\}$ of class $H^s(\Omega) \times H^{s-1}(\Omega)$ in space, $\frac{3}{2} < s \leq 2$, subject to a compatibility condition (c.c.); Theorem 5.2 which yields (less regular) solutions $\{w, w_t\}$ of class $H^s(\Omega) \times H^{s-1}(\Omega)$ in space, $s < \frac{3}{2}$, the maximal regularity where the boundary conditions do not interfere, and thus compatibility conditions are not required; and Theorem 5.3 which provides solutions $\{w, w_t\}$ in the 'energy' space $H^1(\Omega) \times L_2(\Omega)$. It is interesting to point out that in the first theorem, the more traditional regularity theory of the linear problem as expressed by Corollary 3.8 will suffice: instead, in the case of the second and third theorems, crucial use will be made of the sharp regularity theory of Section 3.

Theorem 5.1. [L-S.1] Let $g \in C^3(R, R)$ and $\frac{3}{2} < s \leq 2$; also let $n = \dim\Omega \leq 3$. For each pair of I.C.,

$$w_0 \in H^s(\Omega) \text{ subject to } \frac{\partial w_0}{\partial \nu} = g(w_0|_\Gamma) \text{ on } \Gamma; \quad w_1 \in H^{s-1}(\Omega) = \mathcal{D}(A^{\frac{s-1}{2}}) \quad (5.2)$$

there exists T^+, $0 < T^+ \leq \infty$, such that the corresponding problem (5.1) admits a unique solution satisfying

$$\{w, w_t\} \in C([0, T^+); H^s(\Omega) \times H^{s-1}(\Omega)). \quad (5.3)$$

Theorem 5.2. [L-S.1] Let $g \in C^3(R, R)$ and $n \leq 3$. If $n = 3$, assume further that the second derivative g'' of g be polynomially bounded: there exist constants $c > 0$, $k \geq 2$ such that

$$|g''(r)| \leq c(1 + |r|^{k-2}), \quad \forall r \in R. \quad (5.4)$$

Let $\varepsilon > 0$ be given with $\varepsilon \leq \frac{1}{10}$ if $n \leq 2$; or $\varepsilon \leq 1/(10k)$ if $n = 3$. For each pair of I.C.

$$w_0 \in H^{\frac{3}{2}-\varepsilon}(\Omega), \quad w_1 \in H^{\frac{1}{2}-\varepsilon}(\Omega), \quad (5.5)$$

there exists T^+, $0 < T^+ \leq \infty$, such that the corresponding problem (5.1) admits a unique solution satisfying

(i) $$\{w, w_t\} \in C([0, T^*) : H^{\frac{3}{2}-\varepsilon}(\Omega) \times H^{\frac{1}{2}-\varepsilon}(\Omega)), \qquad (5.6)$$

(ii) $$w|_\Sigma \in L_2(0, T : H^{\frac{1}{2}+\alpha-\varepsilon}(\Gamma)), \qquad (5.7)$$

(iii) $$w_t|_\Sigma \in H^{\alpha-\frac{1}{2}-\varepsilon}(\Sigma), \qquad (5.8)$$

where α is the constant defined by (3.2).

Remark 5.1. Under appropriate structural conditions imposed on the function g, the local solutions of Theorems 5.1 and 5.2 are global solutions as well; i.e., $T^+ = \infty$; see [L-S.1].

The proof of Theorems 5.1 and 5.2 will be sketched in the next two subsections. Instead, the proof of Theorem 5.3 will be (essentially) contained in the proof of the stabilization problem of Theorem 6.2 in Section 6.4.

Theorem 5.3. [Las.2] Let $\dim \Omega = 2$, and let the scalar function g satisfy the conditions (same as (6.23) below),

$$|g''(z)| \leq c|z|^s + d, \qquad \text{for some } s > 0.$$

Then for each pair of I.C. $\{w_0, w_1\} \in H^1(\Omega) \times L_2(\Omega)$, there exists $0 < T^+ \leq \infty$ such that the corresponding problem (5.1) admits a unique solution satisfying

$$\{w, w_t\} \in C([0, T^+]; H^1(\Omega) \times L_2(\Omega)).$$

5.2. Proof of Theorem 5.1.

Let G be the Nemytski operator associated with g:

$$(Gz)(x) = g(z(x)). \qquad (5.9)$$

The proof is based on a fixed point argument.

Step 1. The sought-after solution of (5.1) is a solution of the system

$$\begin{vmatrix} w(t) \\ w_t(t) \end{vmatrix} = \begin{vmatrix} F_1(w, w_t) \\ F_2(w, w_t) \end{vmatrix} \equiv F(w, w_t); \qquad \begin{matrix} \text{(a)} \\ \text{(b)} \end{matrix} \quad (5.10)$$

$$F_1(w, w_t) \equiv C(t)[w_0 - NG(w_0|_\Gamma)] + S(t)w_1 + NG(w_0|_\Gamma) + \int_0^t N[G'(w(\tau)|_\Gamma)(w_t(\tau)|_\Gamma)]d\tau$$
$$- \int_0^t C(t-\tau)N[G'(w(\tau)|_\Gamma)(w_t(\tau)|_\Gamma)]d\tau; \qquad (5.11)$$

$$F_2(w, w_t) \equiv -AS(t)[w_0 - NG(w_0|_\Gamma)] + C(t)w_1 + \int_0^t AS(t-\tau)N[G'(w(\tau)|_\Gamma)(w_t(\tau)|_\Gamma)]d\tau. \qquad (5.12)$$

In fact, by (4.66), (4.68), the abstract model for problem (5.1) is

$$w(t) = C(t)w_0 + S(t)w_1 + \int_0^t AS(t - \tau)NG(w(\tau)|_\Gamma)d\tau. \qquad (5.13)$$

Integrating by parts in t the integral term of (5.13) with $dC(t - \tau)/d\tau = AS(t - \tau)$ and using

$$\int_0^t N\frac{d}{d\tau}[G(w(\tau)|_\Gamma)]d\tau = \int_0^t N[G'(w(\tau)|_\Gamma)(w_t(\tau)|_\Gamma)]d\tau = NG(w(t)|_\Gamma) - N(w_0|_\Gamma), \qquad (5.14)$$

we readily obtain $w(t) = F_1(w, w_t)$, i.e., (5.10a). Next, differentiating (5.10a), we readily obtain $w_t(t) = F_2(w, w_t)$, i.e., (5.10b).

Step 2. Our goal of showing well-posedness on $[0, T]$ is achieved if we can show a contraction fixed point for (5.10) on the space

$$C([0, T]; X_1 \times X_2), \qquad (5.15a)$$

$$X_1 = \{x \in H^s(\Omega) : \|x\|_{H^s(\Omega)} \le R\}; \qquad X_2 = \{x \in H^{s-1}(\Omega); \|x\|_{H^{s-1}(\Omega)} \le R\}, \qquad (5.15b)$$

for suitable $R > 0$ and $T > 0$, and $\frac{3}{2} < s \le 2$.

Step 3. Under the assumptions of Theorem 5.1, the map

$$\{u, v\} \to G'(u)v : H^{s-\frac{1}{2}}(\Gamma) \times H^{s-\frac{3}{2}}(\Gamma) \to H^{s-\frac{3}{2}}(\Gamma) \qquad (5.16)$$

is Lipschitz continuous on bounded sets [L-S.1; Lemma 3.1]. In particular, this result implies that

$$\|G'(y_1|_\Gamma)(\dot{y}_1|_\Gamma) - G'(y_2|_\Gamma)(\dot{y}_2|_\Gamma)\|_{L_2(0,T;H^{s-\frac{3}{2}}(\Gamma))}$$
$$\le c_R\sqrt{T}\{\|y_1 - y_2\|_{C([0,T];X_1)} + \|\dot{y}_1 - \dot{y}_2\|_{C([0,T];X_2)}\}. \qquad (5.17)$$

Step 4. By the (non-sharp) regularity of Corollary 3.8 with $0 \le \theta = 2s - 3 \le 1$ recalling L, L_t in (4.68) we have

$$L : \text{continuous } L_2(0, T; H^{s-\frac{3}{2}}(\Gamma)) \to C([0, T]; H^{s-1}(\Omega)), \qquad (5.18)$$
$$L_t : \text{continuous } L_2(0, T; H^{s-\frac{3}{2}}(\Gamma)) \to C([0, T]; H^{s-2}(\Omega)), \qquad (5.19)$$

and by elliptic regularity, (5.19) yields

$$(A^{-1}L_tv)(t) = \int_0^t C(t-\tau)Nv(\tau)d\tau : \text{continuous } L_2(0, T; H^{s-\frac{3}{2}}(\Gamma)) \to C([0, T]; H^s(\Omega)). \qquad (5.20)$$

Step 5. Recalling (5.2) and the regularity of N in (4.60),

$$w_0 - NG(w_0|_\Gamma) \in \mathcal{D}(A^{s/2}) \subset H^s(\Omega) \qquad (5.21)$$

by [G.1], since $y_0 - NG(y_0|_\Gamma) \in H^s(\Omega)$ and $\frac{\partial}{\partial\nu}[w_0 - NG(w_0|_\Gamma)] = 0$; by (5.21),

$$C(t)[w_0 - NG(w_0|_\Gamma)] + S(t)w_1 \in C([0, T]; \mathcal{D}(A^{s/2})); \qquad (5.22)$$

$$AS(t)[w_0 - NG(w_0|_\Gamma)] + C(t)w_1 \in C([0, T]; \mathcal{D}(A^{\frac{s-1}{2}}) = H^{s-1}(\Omega)). \tag{5.23}$$

Thus, by (5.22) and (5.23), the terms of $F_1(w, w_t)$ and, respectively, $F_2(w, w_t)$ involving the initial data lie in the space (5.15) where we seek a fixed point solution of (5.10).

Step 6. By using (5.17), and respectively (5.17) combined with (5.20), at the level of the two integral terms of F_1 in (5.11), as well as (5.17) combined with (5.18) at the level of the integral term of F_2 in (5.12), we finally see that the map F is contraction mapping and has a fixed point solution in the required space (5.15) for T sufficiently small. $\square$

5.3. Proof of Theorem 5.2.

We again seek to establish a contraction/fixed point argument but this time for the variables $\{w, w_t|_\Gamma\}$ on appropriate spaces.

Step 1. The sought-after solution is a solution of the system

$$\begin{vmatrix} w(t) \\ w_t(t)|_\Gamma \end{vmatrix} = \begin{vmatrix} F_1(w, w_t|_\Gamma) \\ \widetilde{F}_2(w, w_t|_\Gamma) \end{vmatrix} = \widetilde{F}(w, w_t|_\Gamma), \tag{5.24}$$

where F_1 is the same operator (5.11), and $\widetilde{F}_2$ is the restriction of F_2 in (5.12), both viewed now as a function of $\{w, w_t|_\Gamma\}$:

$$\widetilde{F}_2(w, w_t|_\Gamma) = \{F_2(w, w_t|_\Gamma)\}_\Gamma = \{-AS(t)[w_0 - NG(w_0|_\Gamma)] + C(t)w_1$$
$$+ \int_0^t AS(t - \tau)N[G'(w(\tau)|_\Gamma)(w_t(\tau)|_\Gamma)]d\tau\}_\Gamma. \tag{5.25}$$

Our goal is to establish that $\widetilde{F}$ is contraction for suitably small $T > 0$ and suitable $R > 0$ on the space

$$C([0, T]; Y_1) \times L_2(0, T; Y_2) \tag{5.26}$$

$$Y_1 = \{y \in H^{\frac{3}{2}-\varepsilon}(\Omega) : \|y\|_{H^{\frac{3}{2}-\varepsilon}(\Omega)} \leq R\}; \quad Y_2 = \{y \in H^{\alpha-\frac{1}{2}-\varepsilon}(\Gamma); \|y\|_{H^{\alpha-\frac{1}{2}-\varepsilon}(\Gamma)} \leq R\}. \tag{5.27}$$

Step 2. By using Sobolev imbedding, elliptic theory and semigroup theory, one can prove the following [L-S.1].

Lemma 5.4. There exists a function $M(u, v)$ of two variables which is bounded for bounded arguments such that

$$\|[-AS(t)[w_0 - NG(w_0|_\Gamma)] + C(t)w_1]_\Gamma\|_{L_2(0,T;H^{\alpha-\frac{1}{2}-\varepsilon}(\Gamma))}$$
$$\leq M(\|w_0\|_{H^{\frac{3}{2}-\varepsilon}(\Omega)}, \|w_1\|_{H^{\frac{1}{2}-\varepsilon}(\Omega)}); \tag{5.28}$$

$$\|C(t)[w_0 - NG(w_0|_\Gamma)] + S(t)w_1 + NG(w_0|_\Gamma)\|_{C([0,T];H^{\frac{3}{2}-\varepsilon}(\Omega))}$$
$$\leq M(\|w_0\|_{H^{\frac{3}{2}-\varepsilon}(\Omega)}, \|w_1\|_{H^{\frac{1}{2}-\varepsilon}(\Omega)}). \quad \square \tag{5.29}$$

Thus, (5.28) and (5.29) used in (5.11) and (5.25) respectively, say that the terms of F_1 and $\widetilde{F}_2$ involving the initial data lie in the space (5.26) where we seek a fixed point solution of (5.24).

Step 3. As in step 3 of Theorem 5.1 we have: the mapping

$$\{u, v|_\Gamma\} \to G'(u|_\Gamma)v|_\Gamma : C([0, T]; H^{\frac{3}{2}-\varepsilon}(\Omega)) \times L_2(0, T; H^{\alpha-\frac{1}{2}-\varepsilon}(\Gamma)) \to L_2(\Sigma) \quad (5.30)$$

is Lipschitz continuous on bounded sets.

Step 4. By the regularity (4.60) of $N : L_2(\Gamma) \to H^{\frac{3}{2}-\varepsilon}(\Omega)$, we have:

$$\int_0^t C(t - \tau)N[G'(y(\tau)|_\Gamma)(y_t(\tau)|_\Gamma)]d\tau \text{ is a contraction}$$

$$C([0, T]; Y_1) \times L_2(0, T; Y_2) \to C([0, T]; H^{\frac{3}{2}-\varepsilon}(\Omega)) \quad (5.31)$$

for suitably small T. Similarly, for $\int_0^t N[G'(y(\tau)|_\Gamma)(v(\tau)|_\Gamma)]d\tau$ via (5.11), we then conclude that for R sufficiently small, $F_1 : C([0, T]; Y_1) \times L_2(0, T; Y_2) \to C([0, T]; Y_1)$ and

$$\|F_1(u_1, v_1) - F_1(u_2, v_2)\|_{C([0,T];Y_1)} \le c_R\sqrt{T}\{\|u_1 - u_2\|_{C([0,T];Y_1)} + \|v_1 - v_2\|_{L_2(0,T;Y_2)}\} \quad (5.32)$$

so that F_1 is contraction, as required, for T suitably small.

Step 5. It is at the level of proving that $\widetilde{F}_2$ is contraction as well that the sharp regularity theory of Section 3 will be critically used. More precisely, we shall invoke (3.6) of Theorem 3.1:

$$u \in L_2(\Sigma) \to Lu|_\Gamma \in H^{2\alpha-1}(\Sigma), \quad (5.33)$$

along with Sobolev imbedding

$$H^{2\alpha-1}(\Sigma) \subset L_{6/(3-2(\alpha-\frac{1}{2}))}(0, T; H^{\alpha-\frac{1}{2}-\varepsilon}(\Sigma)), \quad (5.34)$$

and Hölder inequality, to finally obtain

$$\|Lg|_\Gamma\|_{L_2(0,T;H^{\alpha-\frac{1}{2}-\varepsilon}(\Gamma))} \le cT^{\frac{\alpha-\frac{1}{2}}{3}}\|Lg|_\Gamma\|_{H^{2\alpha-1}(\Sigma)} \le cT^{\frac{\alpha-\frac{1}{2}}{3}}\|g\|_{L_2(\Sigma)}. \quad (5.35)$$

Step 6. As a consequence of steps 3 through 5, one readily shows

$$\|\{L[G'(u_1)v_1 - G'(u_2)v_2]\}|_\Gamma\|_{L_2(0,T;Y_2)} \le c_r T^{\frac{\alpha-\frac{1}{2}}{3}}\{\|u_1 - u_2\|_{C([0,T];Y_1)} + \quad (5.36)$$
$$\|v_1 - v_2\|_{C([0,T];Y_2)}\},$$

which proves that $\widetilde{F}_2 : C([0, T]; Y_1) \times C[0, T]; Y_2) \to L_2(0, T; Y_2)$ for R sufficiently small and then $\widetilde{F}_2$ is contraction on the required spaces.

Thus we obtain for the fixed point solution that

$$\{w, w_t|_\Gamma\} \in C([0, T]; H^{\frac{3}{2}-\varepsilon}(\Omega)) \times L_2(0, T; H^{\alpha-\frac{1}{2}-\varepsilon}(\Gamma)), \quad (5.37)$$

i.e., part of (5.6) and (5.8).

Step 7. The required regularity for $w_t \in C([0, T]; H^{\frac{1}{2}-\varepsilon}(\Omega))$ as in (5.6) follows directly through steps 3–6 and the regularity of $C(t)$, $S(t)$, N in (4.60) which gives *a-posteriori*

$$G'(w|_\Gamma)w_t|_\Gamma \in L_2(\Gamma), \qquad . \tag{5.38}$$

and from the regularity of L.

For a proof of the additional regularity in (ii) = (5.7) and (iii) = (5.8) we refer to [L-S.1].

Notes. *The proofs of this section make important use, to set up the fixed point arguments, of the explicit abstract model (Section 4) for mixed problems in the integral form, along with the regularity properties of the various ingredients which enter into this representation. This technique of proof has been extensively used by the authors for the solution of various boundary feedback problems for hyperbolic equations, see e.g., [L-T.11]–[L-T.14], and parabolic equations, see e.g., [L-T.15], [L-T.16]. For an application of this technique in the case of fourth-order operators, see [St.1].*

6. Local Exponential Stability of Damped Wave Equations with Semi-linear Boundary Conditions

6.1. Orientation

In this section we consider a wave equation subject to the action of two operators: a friction or damping (linear) operator F depending on the velocity and acting in the interior of Ω; and a nonlinear perturbation operator G (which may model uncertainty of the system) depending on the position and acting on the boundary Γ, see the specific models (6.1) and (6.20) for the Dirichlet and Neumann cases, respectively. It is assumed, as a starting point, that with $G = 0$ the wave equation subject to the operator F is uniformly (exponentially) stable; i.e., that F is a (distributed) stabilizing feedback operator of the (otherwise typically conservative) wave equation. (We may think of F as modelling viscous damping: $\{F(z)\}(x) = \gamma(x)z(x)$, $x \in \Omega$, $\gamma(x) \geq \gamma_0 > 0$ and C^1.) We then address the question: is the new problem subject to F and now also G still locally exponentially stable when G belongs to a sufficiently large class of nonlinear perturbation operators? The answer will be in the affirmative and thus the results presented will fit into the so-called principle of linearized stability: the nonlinearly perturbed system is locally exponentially stable as long as its linearization is uniformly (exponentially) stable. Precise versions of this principle are well known for (finite dimensional) ordinary differential equations as well as for abstract equations acted upon by nonlinear perturbations which are *bounded*. The novelty of the present problems, which introduces genuine additional difficulties over established literature, is that in our case the nonlinear perturbation G acts in the boundary conditions and hence effects the system through *unbounded* operators (AD in the Dirichlet case, AN in the Neumann case, see (4.19) and (4.65) respectively

To overcome these difficulties, we shall show that it is critical to use the regularity results of the linear theory presented in Sections 2 and 3. In what follows, we shall treat the Dirichlet and Neumann case separately.

6.2. Dirichlet Boundary Conditions: Theorem 6.1.

We consider the following wave problem in $w(t, x)$:

$$\begin{cases} w_{tt} + A(x, \partial)w = -F(w_t) & \text{in } (0, \infty) \times \Omega, & (6.1a) \\ w(0, \cdot) = w_0 \in L_2(\Omega) : w_t(0, \cdot) = w_1 \in H^{-1}(\Omega) & \text{in } \Omega, & (6.1b) \\ w|_\Sigma = G(w) & \text{in } (0, \infty) \times \Gamma, & (6.1c) \end{cases}$$

with $A(x, \partial)$ given by (1.4), canonically $A(x, \partial) = -\Delta$. Let A be the resulting realization operator of $A(x, \partial)$, which we take to be self-adjoint and boundedly invertible for simplicity of exposition. We shall make the following

Assumptions.

(i) $\qquad\qquad F :$ continuous on $H^{-1}(\Omega) : \quad F \in \mathcal{L}(H^{-1}(\Omega)) :$ $\qquad\qquad$ (6.2)

(ii) the unique solution, call it $\{\overline{w}, \overline{w}_t\}$ of problem (6.1) with $G = 0$ is uniformly stable on $L_2(\Omega) \times H^{-1}(\Omega)$: there exist $C \geq 1$ and $\omega_0 > 0$ (depending on F) such that

$$|\overline{w}(t)|^2_{L_2(\Omega)} + |\overline{w}_t(t)|^2_{H^{-1}(\Omega)} \leq Ce^{-\omega_0 t}\{|w_0|^2_{L_2(\Omega)} + |w_1|^2_{H^{-1}(\Omega)}\} \qquad (6.3)$$

(equivalently, the semigroup $e^{\mathcal{A}_F t}$, with $\mathcal{A}_F$ as in (6.8) below, is uniformly stable in $L_2(\Omega) \times H^{-1}(\Omega)$);

(iii) G is a nonlinear operator: $L_2(\Omega) \to L_2(\Gamma)$, with Frechet derivative $G'(z)$ satisfying

$$G(0) = 0; \quad |G'(z)|_{\mathcal{L}(L_2(\Omega);L_2(\Gamma))} \to 0 \text{ as } |z|_{L_2(\Omega)} \to 0. \qquad (6.4)$$

The assumptions on G are typical for a perturbation operator that may model uncertainty of the system. The main result is

Theorem 6.1. Assume (i) = (6.2) through (iii) = (6.4). Fix $0 < \omega < \omega_0$. There exists a constant $R > 0$, depending on ω, such that if

$$|w_0|^2_{L_2(\Omega)} + |w_1|^2_{H^{-1}(\Omega)} \leq R, \qquad (6.5)$$

then problem (6.1) admits a unique solution which moreover satisfies the condition: there exist $M_R = M \geq 1$ depending on R such that

$$|w(t)|^2_{L_2(\Omega)} + |w_t(t)|^2_{H^{-1}(\Omega)} \leq Me^{-\omega t}. \qquad (6.6)$$

6.3. Dirichlet Boundary Conditions: Proof of Theorem 6.1.

Step 1. The first-order abstract model of problem (6.1) is

$$\frac{d}{dt}\begin{vmatrix} w \\ w_t \end{vmatrix} = \mathcal{A}_F \begin{vmatrix} w \\ w_t \end{vmatrix} + \mathcal{B}Gw, \qquad (6.7)$$

where $\mathcal{B}$ is the operator in (4.27), the spaces are as in (4.25):

$$Y = Y_1 \times Y_2; \quad Y_1 = L_2(\Omega); \quad Y_2 = H^{-1}(\Omega) = [\mathcal{D}(A^{\frac{1}{2}})]',$$

and

$$\mathcal{A}_F = \begin{vmatrix} 0 & I \\ -A & -F \end{vmatrix}; \quad \mathcal{A}_F^* = \begin{vmatrix} 0 & -I \\ A & -F^* \end{vmatrix}, \qquad (6.8)$$

$\mathcal{A}_F^*$ being the Y-adjoint of $\mathcal{A}_F$, F^* the $[\mathcal{D}(A^{\frac{1}{2}})]'$-adjoint of F. By (6.2), $\mathcal{A}_F$ and $\mathcal{A}_F^*$ generate s.c. semigroups on Y.

Step 2. According to an abstract result of [Las.2], proved by fixed point techniques, the desired conclusion (6.6) of Theorem 6.1 will follow, as soon as we show that

$$B^* e^{\mathcal{A}_F^* t} : \text{ continuous } Y \to L_2(0, T; L_2(\Gamma)) \qquad (6.9)$$

(this extends known results from B bounded to B unbounded).

Step 3. To prove (6.9), we take $y = [y_1, y_2] \in Y$ and consider

$$z_{tt} = -Az - F^* z_t; \quad z(0) = y_1, \quad -z_t(0) = y_2; \quad \begin{vmatrix} z(t) \\ -z_t(t) \end{vmatrix} = e^{\mathcal{A}_F^* t} \begin{vmatrix} y_1 \\ y_2 \end{vmatrix}, \qquad (6.10)$$

whose solution is explicitly

$$z(t) = C(t)y_1 + S(t)y_2 - \int_0^t S(t-\tau)F^* z_t(\tau)d\tau, \qquad (6.11)$$

$$z_t(t) = -AS(t)y_1 + C(t)y_2 - \int_0^t C(t-\tau)F^* z_t(\tau)d\tau, \qquad (6.12)$$

$C(t)$ and $S(t)$ cosine and sine operators of $-A$, where we plainly have

$$|z_t|_{C([0,T];H^{-1}(\Omega))} \le c_T\{|z(0)|_{L_2(\Omega)} + |\dot{z}_t(0)|_{H^{-1}(\Omega)}\}. \qquad (6.13)$$

Step 4. Invoking (4.32) with A self-adjoint, we obtain by (6.10), (6.12),

$$B^* e^{\mathcal{A}_F^* t} y = - D^* z_t(t) = D^* AS(t)y_1 - D^* AC(t)A^{-1}y_2$$
$$+ D^* A \int_0^t C(t-\tau)A^{-1}F^* z_t(\tau)d\tau \qquad (6.14)$$

Recalling the trace regularity results (4.35), (4.36), of Theorem 4.1 we have

$$D^* AS(t)y_1 + D^* AC(t)A^{-1}y_2 \in L_2(0, T; L_2(\Gamma)) \qquad (6.15)$$

continuously in $[y_1, y_2] \in Y$. As to the last term in (6.14), we introduce the following second-order problem,

$$\zeta_{tt} = -A\zeta - A^{-1}F^* z_t; \quad \zeta(0) = \zeta_t(0) = 0, \qquad (6.16)$$

for which we have, recalling (4.16), (4.23),

$$\frac{\partial \zeta_t}{\partial \nu} = -D^* A \int_0^t C(t-\tau)A^{-1}F^* z_t(\tau)d\tau = \frac{d}{dt}D^* AK[-A^{-1}F^* z_t], \qquad (6.17)$$

where in view of (6.13) we have the regularity

$$|A^{-1}F^* z_t|_{C([0,T];H_0^1(\Omega))} \le c_T\{|y_1|_{L_2(\Omega)} + |y_2|_{H^{-1}(\Omega)}\}. \qquad (6.18)$$

Invoking the trace regularity result (4.54) of Theorem 4.3 on (6.17), (6.18) (i.e., (2.24) of Theorem 2.3), we obtain

$$\frac{\partial \zeta}{\partial \nu} = D^* AK[-A^{-1}F^* z_t] \in H^1(\Sigma); \quad \text{a fortiori } \frac{\partial \zeta_t}{\partial \nu} \in L_2(\Sigma). \qquad (6.19)$$

Using (6.15) and (6.17), (6.19) in (6.14) proves (6.9) as desired.

6.4. Neumann Boundary Conditions. Theorem 6.2.

We consider the following wave problem in $w(t, x)$,

$$
\begin{cases}
w_{tt} + A(x, \partial)w = F(w_t) & \text{in } (0, \infty) \times \Omega, & (6.20a) \\
w(o, \cdot))w_0 \in H^1(\Omega) : w_t(0, \cdot) = w_1 \in L_2(\Omega) & \text{in } \Omega, & (6.20b) \\
\frac{\partial w}{\partial \nu}|_\Sigma = G(w) & \text{in } (0, \infty) \times \Gamma, & (6.20c)
\end{cases}
$$

and let A be the realization of $A(x, \partial)$ given by (4.55); canonically $A(x, \partial) = -\Delta + 1$. Again, we shall assume A to be self-adjoint and boundedly invertible. We shall make the following

Assumptions.

(i) $F :$ continuous on $L_2(\Omega) :$ $F \in \mathcal{L}(L_2(\Omega)) :$ (6.21)

(ii)   the unique solution, call it $\{\overline{w}, \overline{w}_t\}$, of problem (6.1) with $G = 0$ is uniformly stable on $H^1(\Omega) \times L_2(\Omega)$: there exist $c \geq 1$ and $\omega_0 > 0$ (depending on F) such that

$$
|\overline{w}(t)|^2_{H^1(\Omega)} + |\overline{w}_t(t)|^2_{L_2(\Omega)} \leq Ce^{-\omega_0 t}\{|w_0|^2_{H^1(\Omega)} + |w_1|^2_{L_2(\Omega)}\} \tag{6.22}
$$

(equivalently, the semigroup $e^{A_F t}$ below is uniformly stable in $H^1(\Omega) \times L_2(\Omega)$).

(iii) G is a Nemytski operator defined by

$$
\left.\begin{array}{l}
(Gz)(x) = g(z(x)|_\Gamma), \quad x \in \Omega, \text{ where } g \in C^3(R, R) \\
\text{is a scalar function satisfying } g(0) = g'(0) = 0; \\
|g''(z)| \leq C|z|^s + d, \text{ for some } s > 0.
\end{array}\right\} \tag{6.23}
$$

Remark 6.1. With A_F^* and B^* the operators in (6.8), (4.73), in the present Neumann case with $Y = H^1(\Omega) \times L_2(\Omega)$ the regularity result $B^* e^{A_F^* t}$: continuous $Y \to L_2(0, T; L_2(\Gamma))$ — present version of statement (6.9) in the Dirichlet case — holds true <u>only if</u> $\dim \Omega = 1$ and fails (even with $F = 0$) if $\dim \Omega \geq 2$. These facts are related, by transposition or duality, to the content of Remark 3.1(iv), (3.21), (3.22). Therefore, in this section, we take $\dim \Omega = 2$ and accordingly an approach different from the one in the Dirichlet case of Section 6.2 will be followed. It will be, at any rate, critically dependent on the sharp regularity results of Section 3 for the linear problem. This explains why assumption (6.23) on G now is of different nature than assumption (6.4) on G in the Dirichlet case. □

Theorem 6.2. Let $\dim \Omega = 2$. Assume (i) = (6.21) through (iii) = (6.23). Fix $0 < \omega < \omega_0$. Then there exists a constant $R > 0$ (depending on ω) such that if

$$
|w_0|^2_{H^1(\Omega)} + |w_1|^2_{L_2(\Omega)} \leq R, \tag{6.24}
$$

then problem (6.20) admits a unique solution which moreover satisfies the condition: there exist $M \geq 1$ depending on R such that

$$
|w(t)|^2_{H^1(\Omega)} + |w_t(t)|^2_{L_2(\Omega)} \leq Me^{-\omega t}. \tag{6.25}
$$

6.5. Neumann Boundary Conditions. Proof of Theorem 6.2

Step 1. Problem (6.1) admits the following semigroup form ((4.76)):

$$\begin{vmatrix} w(t) \\ w_t(t) \end{vmatrix} = e^{\mathcal{A}_F t} \begin{vmatrix} w_0 \\ w_1 \end{vmatrix} + \int_0^t e^{\mathcal{A}_F(t-\tau)} \mathcal{B} G w(\tau) d\tau, \tag{6.26}$$

where $\mathcal{A}_F$ is the same operator as in (6.8) except that A is now defined by (4.55), and $\mathcal{B}$ is defined by (4.73). The space Y is now $Y = H^1(\Omega) \times L_2(\Omega)$. To prove Theorem 6.2, we shall seek a fixed point solution of the integral equation

$$y(t) = e^{(\mathcal{A}_F + \omega I)t} y_0 + \int_0^t e^{(\mathcal{A}_F + \omega I)(t-\tau)} \mathcal{B}[e^{\omega\tau} G(e^{-\omega\tau} y_1(\tau))] d\tau, \tag{6.27}$$

where $y(t) = [y_1(t), y_2(t)] = e^{\omega t}[w(t), w_t(t)]$, $0 < \omega < \omega_0$, in the space

$$E \equiv C([0,\infty]; H^1(\Omega) \times L_2(\Omega)) \cap L_2(0,\infty; H^1(\Omega) \times L_2(\Omega)) \tag{6.28}$$

by showing that the map (right hand side of (6.27))

$$\mathcal{M}v \equiv e^{(\mathcal{A}_F + \omega t)} y_0 + \{\mathcal{L}_{F,\omega}(e^{\omega\tau} G(e^{-\omega\tau} v_1(\tau))\}(t), \tag{6.29}$$

$$(\mathcal{L}_{F,\omega}u)(t) = \int_0^t e^{(\mathcal{A}_F + \omega I)(t-\tau)} \mathcal{B}u(\tau) d\tau \tag{6.30}$$

is a contraction on the closed ball $B(0,R)$ of radius R of the space E in (6.28). To accomplish this, it will be critical to use the sharp regularity results of Section 3 for the linear problem, combined with the following

Step 2. Lemma 6.3. [Las.2] With $\Sigma = (0,\infty) \times \Gamma$, define the space

$$X = H^{\frac{1}{2}}(\Sigma) \cap C([0,\infty]; L_p(\Gamma)) \cap L_2(0,\infty; L_p(\Gamma)), \tag{6.31}$$

$p = 2(s+1)/\varepsilon$, $0 < \varepsilon < \frac{1}{2}$, s as in (6.23). Then the operator G defined by (6.23) is Lipschitz continuous: $X \to H^{\frac{1}{2}-\varepsilon}(\Sigma)$, with Lipschitz constant $\ell(|z|_X) \to 0$ as $|z|_X \to 0$. $\square$

Step 3. Invoking the regularity result in (3.27) of Corollary 3.4, we have that for $F = 0$ and any $0 < T < \infty$;

$$\mathcal{L}_{0,\omega} : \text{ continuous } H^{1-\alpha}(\Sigma_T) \to C([0,T]; H^1(\Omega) \times L_2(\Omega)), \tag{6.32}$$

where α is the constant defined by (3.2).

Step 4. The regularity in (6.32) is preserved under addition of a bounded F as in (6.21):

$$\mathcal{L}_{F,\omega} : \text{ continuous } H^{1-\alpha}(\Sigma_T) \to C([0,T]; H^1(\Omega) \times L_2(\Omega)), \tag{6.33}$$

Step 5. Exploiting the exponential decay of the semigroup $\exp(\mathcal{A}_F t)$ as per assumption (ii) = (6.22), one extends (6.33) to $T = \infty$.

Lemma 6.4. With $Y = H^1(\Omega) \times L_2(\Omega)$,

$$\mathcal{L}_{F,\omega} : \text{ continuous } H^{1-\alpha}((0,\infty) \times \Gamma) \to C([0,\infty]; Y) \cap L_2(0,\infty; Y). \tag{6.34}$$

Step 6. We can finally conclude that the map $\mathcal{M}$ in (6.29) is a contraction as desired. To this end, we use $H^1(\Omega) \subset L^p(\Omega)$, $1 \le p < \infty$ (dim $\Omega = 2$); Lemma 6.3 applied to the nonlinear term $G(e^{-\omega\tau}y_1)$ and Lemma 6.4 with $\frac{1}{2} - \varepsilon$ of Lemma 6.3 equal to $1 - \alpha$ of Lemma 6.4: $\frac{1}{2} - \varepsilon = 1 - \alpha$. This is legal since $1 - \alpha < \frac{1}{2}$ by (3.2)!

Remark 6.2. The last step 6 shows that it is critical to the argument to have the constant $\alpha > \frac{1}{2}$, i.e., to use the sharp regularity theory. Instead, the earlier theory as in [L-M.1], [M.1] where $\alpha = \frac{1}{2}$ is inadequate.

Remark 6.3. The proof of Theorem 6.2 contains the proof of the local regularity result in Theorem 5.3 (there is no need now to multiply (6.26) by $e^{\omega t}$ or to take $g(0) = g'(0) = 0$, as contraction can be obtained by taking T small).

7. Exact Controllability of Semi-linear Hyperbolic Problems

The notion of exact controllability for second-order hyperbolic mixed problems (as well as for plate equations, not necessarily hyperbolic), besides being of interest in itself, has crucial bearing on the issue of verifying the Finite Cost Condition (8.7) which is needed in the study of optimal control problems with quadratic functional (8.3) and $T = \infty$, and related Algebraic Riccati Equations. In fact, we shall appeal to exact controllability of linear hyperbolic dynamics of Dirichlet type in Section 8.5 precisely for this purpose. In this section we shall present, following [L-T.15], recent results on exact controllability of semi-linear hyperbolic problems of both Dirichlet and Neumann type, where the non-linearity may occur either on the equation, or else in the boundary conditions. In line with the main thrust of the second part of this paper, the results on exact controllability of semi-linear problems to be presented here crucially rely on the regularity properties of the hyperbolic dynamics given in Sections 2 and 3. We shall treat separately the Dirichlet and the Neumann cases.

7.1. Semi-linear Wave Equation of Neumann Type. Exact Controllability

In Ω, with dim $\Omega \ge 2$, we consider the problem in $w(t, x)$,

$$\begin{cases} w_{tt} = \Delta w & \text{in } Q = (0, T) \times \Omega; & (7.1a) \\ w(0, \cdot) = w_0 : w_t(0, \cdot) = w_1 & \text{in } \Omega; & (7.1b) \\ w|_{\Sigma_0} = 0 & \text{in } \Sigma_0 = (0, T] \times \Gamma_0; & (7.1c) \\ \frac{\partial w}{\partial \nu}|_{\Sigma_1} = g(w|_{\Sigma_1}) + u & \text{in } \Sigma_1 = (0, T] \times \Gamma_1; & (7.1d) \end{cases}$$

$\Gamma_0 \cup \Gamma_1 = \Gamma$, $\bar{\Gamma}_0 \cap \bar{\Gamma}_1 = \emptyset$. Moreover, on the basis of the sharp regularity results of Section 3, Theorem 3.1, the scalar function g is assumed to satisfy the following assumption when dim $\Omega \ge 2$:

$$g : \text{continuous } H^{2\alpha-1}(\Sigma) \to L_2(\Sigma), \tag{7.2}$$

where α is defined by (3.2); i.e., with $\varepsilon > 0$ arbitrary

$$2\alpha - 1 = \begin{cases} \frac{1}{3} - \epsilon, & \text{if } \Omega \text{ is a general smooth domain}; & (7.3a) \\ \frac{1}{4}, & \text{if } \Omega \text{ is a parallelepiped}; & (7.3b) \\ \frac{1}{3}, & \text{if } \Omega \text{ is a sphere}; & (7.3c) \end{cases}$$

Theorem 7.1. With reference to problem (7.1) subject to assumption (7.2), let μ be an $L_2(\Sigma_1)$-control function that steers the origin $\{0, 0\}$ to the state $\{v(T, \cdot), v_t(T, \cdot)\} \in H_{\Gamma_0}^1(\Omega) \times L_2(\Omega)$ at time T, along the solution of the linear problem

$$
\begin{cases}
v_{tt} = \Delta v & \text{in } Q = (0, T) \times \Omega; & (7.4a) \\
v(0, \cdot) = v_t(0, \cdot) = 0 & \text{in } \Omega; & (7.4b) \\
v|_{\Sigma_0} \equiv 0 & \text{in } \Sigma_0 = (0, T] \times \Gamma_0; & (7.4c) \\
\frac{\partial v}{\partial \nu}|_{\Sigma_1} = \mu \in L_2(\Sigma_1) & \text{in } \Sigma_1 = (0, T] \times \Gamma_1. & (7.4d)
\end{cases}
$$

Then the control function

$$
u = \mu - g(v|_{\Sigma_1}) \in L_2(\Sigma_1) \tag{7.5}
$$

used in (7.1c) of the linear problem (7.1) with $w_0 = w_1 = 0$ produces the same solution

$$
w(t, \cdot) \equiv v(t, \cdot), \quad w_t(t, \cdot) \equiv v_t(t, \cdot), \quad 0 \le t \le T. \tag{7.6}
$$

In particular, $\{w(T, \cdot), w_t(T, \cdot)\} = \{v(T, \cdot), v_t(T, \cdot)\}$. Thus, in particular, problem (7.1) is exactly controllable on $H_{\Gamma_0}^1(\Omega) \times L_2(\Omega)$ at time $T > 0$, whenever problem (7.4) is. For exact controllability results for problem (7.4) see the recent direct approaches in [Lio.3], [L-T.18], [B-L-R], which followed the original results via uniform stabilization in [C.1], [Lag.1].

Remark 7.1. Notice that assumption (7.2) on g is satisfied if for instance

$$
|g(s)| \le a + b|s|^k, \quad k < \frac{n}{n - 2(2\alpha - 1)} \tag{7.7}
$$

Hence, g may be superlinear.

Remark 7.2. Theorem 7.1 with $2\alpha - 1 = \frac{1}{5} - \varepsilon$ in a general smooth domain Ω with $\dim \Omega \ge 2$, remains true if $(-\Delta)$ is replaced by $A(x, \partial)$ defined in (1.4), and the normal derivative is replaced by the corresponding co-normal derivative.

Proof of Theorem 7.1. The crux of the proof is precisely the trace regularity result (3.6) of Theorem 3.1 as applied to problem (7.4): thus $\mu \in L_2(\Sigma_1)$ as assumed implies $v|_{\Sigma_1} \in H^{2\alpha-1}(\Sigma_1)$. By assumption (7.2) on g, we then have that $g(v|_{\Sigma_1}) \in L_2(\Sigma_1)$. We next define the function u as in (7.5). Then $u \in L_2(\Sigma_1)$ as desired. We finally note that the general question of well-posedness of problem (7.1) under the assumption (7.7) with a general $u \in L_2(\Sigma)$ is handled by the techniques presented in Section 5.

We now turn to another semilinear problem for second-order hyperbolic equations of Neumann type, where again the non-linearity and the control function appear as additive terms, this time on the interior rather than on the boundary. Consider the problem

$$
\begin{cases}
w_{tt} = \Delta w + h(w|_\Sigma) + u & \text{in } Q; & (7.8a) \\
w(0, \cdot) = w_0; w_t(0, \cdot) = w_1 & \text{in } \Omega; & (7.8b) \\
\frac{\partial w}{\partial \nu}|_{\Sigma_1} \equiv 0 & \text{in } \Sigma; & (7.8c)
\end{cases}
$$

where the scalar function h now satisfies the following assumption when $\dim \Omega \ge 2$:

$$
h : \text{ continuous } H^\beta(\Sigma) \to L_2(Q) \tag{7.9}
$$

with constant β defined in (3.2).

Theorem 7.2. With reference to problem (7.8) subject to assumption (7.9), let μ be the $L_2(Q)$-control function that steers the origin $\{0, 0\}$ to the state $\{v(T, \cdot), v_t(T, \cdot)\} \in H^1(\Omega) \times L_2(\Omega)$ at time T, along the solution of the linear problem

$$\begin{cases} v_{tt} = \Delta v + \mu & \text{in } Q; & \text{(7.10a)} \\ v(0, \cdot) = v_t(0, \cdot) = 0 & \text{in } \Omega; & \text{(7.10b)} \\ \frac{\partial v}{\partial v}|_{\Sigma} = 0 & \text{in } \Sigma. & \text{(7.10c)} \end{cases}$$

Then the control function

$$u = \mu - h(v|_{\Sigma}) \in L_2(Q) \tag{7.11}$$

used in (7.8a) of the nonlinear problem (7.8) with $w_0 = w_1 = 0$ produces the same solution $w(t, \cdot) \equiv v(t, \cdot)$, $w_t(t, \cdot) \equiv v_t(t, \cdot)$, $0 \le t \le T$. Thus, in particular, problem (7.8) is exactly controllable on $H^1(\Omega) \times L_2(\Omega)$ at any time $T > 0$, since so is problem (7.10) [Tr.5].

Proof of Theorem 7.2. The proof is similar to that of Theorem 7.1, except that it now uses the trace regularity (3.15) of Theorem 3.3, as applied to problem (7.10): $\mu \in L_2(Q)$, as assumed, yields $v|_{\Sigma} \in H^\beta(\Sigma)$: hence (7.11) follows via (7.9).

7.2. Semi-linear Wave Equation of Dirichlet Type. Exact Controllability

With $\dim \Omega \ge 1$, we now consider

$$\begin{cases} w_{tt} = \Delta w + f(w) & \text{in } Q = (0, T] \times \Omega; & \text{(7.12a)} \\ w(0, \cdot) = w_0; w_t(0, \cdot) = w_1 & \text{in } \Omega; & \text{(7.12b)} \\ w|_{\Sigma_0} = 0 & \text{in } \Sigma_0 = (0, T] \times \Gamma_0; & \text{(7.12c)} \\ w|_{\Sigma_1} = u & \text{in } \Sigma_1 = (0, T] \times \Gamma_1; & \text{(7.12d)} \end{cases}$$

with control action u based on Γ_1, $\Gamma_0 \cup \Gamma_1 = \Gamma$, $\overline{\Gamma}_0 \cap \overline{\Gamma}_1 = \emptyset$. The assumption on the non-linear function f is

$$f \in W^{1,\infty}(\mathbb{R}); \quad \text{hence} \quad |f'(r)| \le c; \quad \text{a.e. } r \in \mathbb{R}, \tag{7.13}$$

i.e., f is absolutely continuous $\mathbb{R} \to \mathbb{R}$ with first derivative f' a.e. which is uniformly bounded in $\mathbb{R}$ a.e. The linear problem (7.12) with $f \equiv 0$ is exactly controllable over $[0, T]$ on the space $L_2(\Omega) \times H^{-1}(\Omega)$ (of optimal regularity as in Theorem 2.1) within the class of controls $u \in' L_2(0, T; L_2(\Gamma_1))$, provided $T > 0$ is sufficiently large and under some geometrical conditions on $\{\Omega, \Gamma_0, \Gamma_1\}$ [H.1], [Lio.3], [T.5], the most general ones being in [B-L-R]; equivalently, on the space $H_0^1(\Omega) \times L_2(\Omega)$ within the class of controls $H_0^1(0, T; L_2(\Gamma_1))$ [L-T.17]; in turn, equivalently, on the space $H_0^\gamma(\Omega) \times H^{\gamma-1}(\Omega)$, $0 < \gamma < 1$, $\gamma \neq \frac{1}{2}$ (respect. $H_{00}^{\frac{1}{2}}(\Omega) \times [H_{00}^{\frac{1}{2}}(\Omega)]'$ if $\gamma = \frac{1}{2}$) within the class of controls $u \in H_0^\gamma(0, T; L_2(\Gamma_1))$ (respect. $u \in H_{00}^{\frac{1}{2}}(0, T; L_2(\Gamma_1))$ if $\gamma = \frac{1}{2}$). The next result extends the same exact controllability property to the semi-linear problem (7.12) subject to assumption (7.13) over the same time intervals and over the same Γ_1.

Theorem 7.3. Let Σ_1 be such that the exact controllability of the linear problem (7.12) with $f \equiv 0$ holds true in any one of the equivalent statements above. Let f satisfy (7.13). Then a similar exact controllability result holds true for the original problem (7.12) with the same Σ_1: for any pair $\{w_0, w_1\} \in H_0^\gamma(\Omega) \times H^{\gamma-1}(\Omega)$, $0 \le \gamma \le 1$, $\gamma \neq \frac{1}{2}$

(respect. $\{w_0, w_1\} \in H_{00}^{\frac{1}{2}}(\Omega) \times [H_{00}^{\frac{1}{2}}(\Omega)]'$ for $\gamma = \frac{1}{2}$), there exists a suitable control function $u \in H_0^\gamma(0, T; L_2(\Gamma_1))$ (respect. $u \in H_{00}^{\frac{1}{2}}(0, T; L_2(\Gamma_1))$) for $\gamma = \frac{1}{2}$ such that the corresponding solution of problem (7.12) satisfies $w(T, \cdot) = w_t(T, \cdot) = 0$. $\square$

The weaker statement of Theorem 7.3 in the open range $0 < \gamma < 1$ was originally given in [Z.1], whose methods could not handle (as explicitly pointed out in [Z.1]) the end cases $\gamma = 0$ and $\gamma = 1$, which are the cases of main interest. This required to look for a different proof, capable to handle also the most demanding cases $\gamma = 0$ and $\gamma = 1$. Such proof was given in [L-T.17] in fact first for an abstract system (Theorem 7.4 below), intended to model both hyperbolic and plate problems, and then specialized to various such dynamics. In subsection 7.4, we shall verify all the abstract assumptions of Theorem 7.4 in the case of the Dirichlet problem (7.12): in doing so, we shall crucially rely on the regularity properties of Section 3 for the corresponding linear problem.

7.3. An Exact Controllability Result for Abstract Semi-linear Systems. Theorem 7.4.

We let Y and U be two Hilbert spaces. The basic operator model is the abstract equation

$$\dot{y} = \mathcal{A}y + \mathcal{F}(y) + \mathcal{B}u, \qquad y(0) = y_0 \in Y, \qquad (7.14)$$

to be interpreted as specified below. Standing assumptions on (7.14) are:

(i)   that $\mathcal{A}$ generates a s.c. semigroup $e^{\mathcal{A}t}$ on Y;
(ii) $\mathcal{B}$ satisfies $\mathcal{A}^{-1}\mathcal{B} \in \mathcal{L}(U, Y)$;
(iii) $\mathcal{F}$ is a nonlinear operator $Y \to Y$, continuous on Y, with Frechet derivative $\mathcal{F}'[y] \in \mathcal{L}(Y)$ at the point $y \in Y$, satisfying

$$|\mathcal{F}'[y]|_{\mathcal{L}(Y)} \le \text{const, uniformly in } y \in Y. \qquad (7.15)$$

Instead of the differential version (7.14) on, say, $[\mathcal{D}(\mathcal{A}^*)]'$, we shall consider its variation of parameter version

$$y(t) = e^{\mathcal{A}t}y_0 + (\mathcal{L}u)(t) + (RFy)(t), \qquad (7.16)$$

$$y(T) = e^{\mathcal{A}T}y_0 + \mathcal{L}_T u + R_T Fy, \qquad (7.17)$$

$$(Fy)(t) = \mathcal{F}(y(t)), \qquad (7.18)$$

to be interpreted under some minimal requirement of well-posedness as follows. There exist two Hilbert spaces:

a Hilbert space $\tilde{\mathcal{U}}_T$, based on $[0, T] \times U$, dense in $L_2(0, T; U)$, $\qquad (7.19)$

and

a Hilbert space $H \subset Y$ (H will be the space of exact controllability at $t = T$), $\qquad (7.20)$

such that

$$(\mathcal{L}u)(t) = \int_0^t e^{A(t-\tau)} \mathcal{B}u(\tau) d\tau : \text{ continuous } \widetilde{\mathcal{U}}_T \to C([0,T];Y); \quad (7.21)$$

$$\mathcal{L}_T u = \int_0^T e^{A(T-\tau)} \mathcal{B}u(\tau) d\tau : \text{ continuous } \widetilde{\mathcal{U}}_T \to H. \quad (7.22)$$

Moreover, in (7.16), (7.17), we have set

$$(Rg)(t) = \int_0^t e^{A(t-\tau)} g(\tau) d\tau : \text{ continuous } L_1(0,T;Y) \to C([0,T];Y); (7.23)$$

$$R_T g = \int_0^T e^{A(T-\tau)} g(t) dt : \text{ continuous } L_1(0,T;Y) \to Y. \quad (7.24)$$

By (7.15), a fixed point solution $y \in C([0,T];Y)$ of (7.16) exists for $u \in \widetilde{\mathcal{U}}_T$. The space $\widetilde{\mathcal{U}}_T$ is invoked only for the well-posedness of (7.14) and will not be needed subsequently. To state the abstract exact controllability result on the space H, we need to introduce two more operators:

$$K[\eta] = R\mathcal{F}'[\eta]; \quad (K[\eta]g)(t) = \int_0^t e^{A(t-\tau)} \mathcal{F}'[\eta]g(\tau) d\tau; \quad (7.25)$$

$$K_T[\eta] = R_T \mathcal{F}'[\eta]; \quad K_T[\eta]g = \int_0^T e^{A(T-t)} \mathcal{F}'[\eta]g(t) dt; \quad (7.26)$$

Exact controllability problem on the space H, at time T, within the class of $\mathcal{U}_T$-controls

Our aim is to obtain an exact controllability result for the dynamics (7.14), i.e., for (7.16), on the state space H, at time T, within the class of $\mathcal{U}_T$-controls, where $\mathcal{U}_T$ is a Hilbert space $\mathcal{U}_T \subset \widetilde{\mathcal{U}}_T$. This means that: given $y_0 \in H$ (respect. $y_t \in H$), we seek, if possible, $u \in \mathcal{U}_T$ such that the corresponding solution of (7.16), (7.17) (respect. with initial condition $y_0 = 0$), satisfies $y(T) = 0$ (respect. $y(T) = y_T$). The two formulations are equivalent in the case of time reversible dynamics such as problem (7.12) (and plate problems as well).

Assumptions and statement of abstract result.

We shall make throughout two sets of assumptions: structural assumptions (A.1) through (A.5) on the operators describing model (7.14); and controllability asumptions (C.1), (C.2) on the linear and linearized versions of problem (7.14). All these assumptions are (perhaps despite their first impression) quite natural for hyperbolic (as well as plate) equations, see [L-T.17]. They will all be verified to hold true in the case of the Dirichlet problem (7.12) in the next subsection.

Structural Assumptions (A1)–(A5)

(A.1): **Assumption on the operator $\mathcal{L}$ defined by (7.21).** There exists a Hilbert space $\mathcal{E}_T$

$$\mathcal{E}_T \supset C([0,T];Y) \quad (7.27)$$

such that

$$\mathcal{L} : \mathcal{U}_T \to \mathcal{E}_T \text{ is continuous} \tag{7.28}$$

(in applications, there is much flexibility in the choice of $\mathcal{E}_T$). Moreover,

$$\text{either} \quad \mathcal{L} : \mathcal{U}_T \to \mathcal{E}_T \text{ compact,} \tag{7.29}$$

$$\text{or else} \quad R_T : \bigcup_{\eta \in \mathcal{E}_T} \mathcal{F}'[\eta]g \to H \text{ is compact, for each } g \in \mathcal{E}_T \text{ fixed.} \tag{7.30}$$

(A.2): Assumptions on the family $K[\eta] = R\mathcal{F}'[\eta]$ defined by (7.25)

Assumption (a). The following family of operators is collectively compact in the parameter $\eta \in \mathcal{E}_T$:

$$K[\eta] = R\mathcal{F}'[\eta] : \mathcal{E}_T \to \mathcal{E}_T. \tag{7.31}$$

This means, explicitly, that the following two properties hold true:

(a$_1$)   for each $\eta \in \mathcal{E}_T$, the operator $K[\eta] : \mathcal{E}_T \to \mathcal{E}_T$ is compact;   (7.32)

(a$_2$)   the set union $\bigcup_{\eta \in \mathcal{E}_T} K[\eta]$ (unit ball of $\mathcal{E}_T$) is a pre-compact set in $\mathcal{E}_T$,   (7.33)

where the union of the image under $K[\eta]$ of the unit ball in $\mathcal{E}_T$ is taken over all η running in $\mathcal{E}_T$.

Assumption (b). For any sequence $\eta_n \in \mathcal{E}_T$, $n = 1, 2, \ldots$, one can extract a subsequence η_{n_k}, $k = 1, 2, \ldots$ such that

$$K[\eta_{n_k}] = R\mathcal{F}'[\eta_{n_k}] \to K^0 \equiv R\mathcal{F}_0 \text{ strongly in } \mathcal{E}_T \tag{7.34}$$

for a suitable operator $\mathcal{F}_0 \in L(Y)$, which may depend on the subsequence.

(A.3) Assumption on the family $K_T[\eta] = R_T\mathcal{F}'[\eta]$ defined by (7.26). The family of operators

$$K_T[\eta] = R_T\mathcal{F}'[\eta] : \mathcal{E}_T \to H \tag{7.35}$$

has the property that: for any sequence $\eta_n \in \mathcal{E}_T$, $n = 1, 2, \ldots$, one can extract a subsequence η_{n_k}, $k = 1, 2, \ldots$ such that

$$K_T[\eta_{n_k}] = R_T\mathcal{F}'[\eta_{n_k}] \to \text{ some } K_T^0 = R_T\mathcal{F}_0 \text{ weakly in } H \text{ from } \mathcal{E}_T, \tag{7.36}$$

i.e.,

$$(K_T[\eta_{n_k}]g, h)_H \to (K_T^0 g, h)_H, \qquad \forall \, g \in \mathcal{E}_T, \quad \forall \, h \in H$$

for a suitable operator $\mathcal{F}_0 \in L(Y)$, which may depend on the subsequence.

The above assumption is needed in the first alternative of (A.1), i.e., when $\mathcal{L}$ is compact as stated by (7.29). On the other hand, in the second alternative of (A.1), i.e., when R_T is compact as stated by (7.30), then the convergence in (7.36) is strong in H in view of (7.15).

(A.4): **Assumption on the trajectory $(R\mathcal{F}(0))(t)$, $0 \le t \le T$.** We have

$$\text{trajectory } \{(R\mathcal{F}(0))(t), \ 0 \le t \le T\} \subset \text{ compact set of } \mathcal{E}_T. \tag{7.37}$$

(A5.): **Assumption on the point $R_T\mathcal{F}(0)$.** We have

$$R_T\mathcal{F}(0) \in H. \tag{7.38}$$

Controllability Assumptions (C.1), (C.2).

(C.1) (exact controllability from the origin on the space H, at time T, of the linear problem (7.14) with $\mathcal{F} \equiv 0$, $y_0 = 0$, within the class of $\mathcal{U}_T$-controls):

$$\mathcal{L}_T : \mathcal{U}_T \to H \text{ is surjective (onto).} \tag{7.39}$$

(C.2) (approximate controllability from the origin of the linearized problem, and its limit version in the sense of (7.34), (7.36)):

(a) for each fixed $\eta \in \mathcal{E}_T$, the map $M_T[\eta]$ defined by

$$u \to M_T[\eta]u = \{\mathcal{L}_T + K_T[\eta](I - K[\eta])^{-1}\mathcal{L}\}u : \mathcal{U}_T \to H \tag{7.40}$$

has range dense in H, in the topology of H. (We note that the inverse of (7.40) is well defined as a consequence of assumption (A.2)(a$_1$)=(7.32).)

(b) Also, let K^0 and K_T^0 be any of the limit operators obtained in (7.34) and (7.36). We likewise assume that the operator

$$M_T^0 = \mathcal{L}_T + K_T^0(I - K^0)^{-1}\mathcal{L} : \mathcal{U}_T \to H \tag{7.41}$$

has range dense in H, in the topology of H. (It can be shown that the inverse in (7.41) is likewise well defined, since K^0 is compact.) An equivalent formulation of (7.40) and (7.41) is that the Hilbert space adjoint operator $M_T^* : H \to \mathcal{U}_T$

$$(M_T u, y)_H = (u, M_T^* y)_{\mathcal{U}_T} \tag{7.42}$$

with M_T either $M_T[\eta]$ or M_T^0, has trivial null space in H:

$$\text{null space } \{M_T^*[\eta]\} = \text{ null space } \{\mathcal{L}_T^* + \mathcal{L}^*(I - K^*[\eta])^{-1}K_T^*[\eta]\} = \{0\} \text{ in } H; \tag{7.43}$$

$$\text{null space } \{(M_T^0)^*\} = \text{ null space } \{\mathcal{L}_T^* + \mathcal{L}^*(I - (K^0)^*)^{-1}(K_T^0)^*\} = \{0\} \text{ in } H, \tag{7.44}$$

Main exact controllability result. We can now state

Theorem 7.4. Assume (A.1) through (A.5) and (C.1), (C.2). Then for any $y_T \in H$, there exists $u \in \mathcal{U}_T$ such that the corresponding solution of (7.14) (or (7.16), (7.17)) with $y_0 = 0$ satisfies $y(T) = y_T$.

The above abstract theorem applies not only to the hyperbolic problem of Dirichlet type (7.12), as it will be demonstrated in the next subsection, but to plate equations as well (see [L-T.17]); see also *Notes* at the end.

7.4. Application of Theorem 7.4 to the Semi-linear Wave Problem (7.12) with Dirichlet Control

We return to problem (7.12) subject to (7.13). We shall now verify that all assumptions of the abstract Theorem 7.4 are fulfilled in this case, and in natural function spaces in fact: to this end, we shall make crucial use of the regularity results of Section 2 for the linear version of problem (7.12). After this verification, Theorem 7.3 will then be nothing but a specialization of the abstract Theorem 7.4. Below we shall verify the case $\gamma = 1$ and $\gamma = 0$ in the statement of Theorem 7.3, with $\Gamma_1 = \Gamma$ and $\Gamma_0 = \emptyset$.

Abstract setting for problem (7.12). For $\gamma = 1$, and with reference to (7.12), we select the following spaces for the abstract framework of Theorem 7.4,

$$Y = L_2(\Omega) \times H^{-1}(\Omega); \qquad H = H_0^1(\Omega) \times L_2(\Omega); \tag{7.45}$$

$$U = L_2(\Gamma); \qquad \mathcal{U}_T = H_0^1(0, T; L_2(\Gamma)); \tag{7.46}$$

$$\mathcal{E}_T = L_2(0, T; L_2(\Omega) \times H^{-1}(\Omega)). \tag{7.47}$$

The operators $\mathcal{A}$, $\mathcal{B}$ occurring in the abstract model (7.14) in the case of problem (7.12) are defined by (4.26), (4.27). The operator $\mathcal{F}$ in (7.14) is defined by

$$\mathcal{F}(y) = \begin{vmatrix} 0 \\ f(y_1(\cdot)) \end{vmatrix}; \quad \mathcal{F}'[\eta]y = \begin{vmatrix} 0 \\ f'(\eta_1(\cdot))y_1(\cdot) \end{vmatrix}; \quad \mathcal{F}(0) = \begin{vmatrix} 0 \\ f(0) \end{vmatrix}. \tag{7.48}$$

Verification of Assumption (A.1), Alternative (7.29)

Proposition 7.5. In the case of problem (7.12), the operator $\mathcal{L}$ in (4.30) satisfies

$$\mathcal{L} : \mathcal{U}_T \equiv H_0^1(0, T; L_2(\Gamma)) \to \mathcal{E}_T \equiv L_2(0, T; L_2(\Omega) \times H^{-1}(\Omega)) \text{ is compact.} \tag{7.49}$$

Proof. We use Aubin's Compactness Lemma. We have

$$\mathcal{L} : \text{ continuous } H_0^1(0, T; L_2(\Gamma)) \to C([0, T]; H^{\frac{1}{2}}(\Omega) \times L_2(\Omega)), \tag{7.50}$$

see Eq. (2.25), and then it suffices to show that

$$\frac{d\mathcal{L}}{dt} : \text{ continuous } H_0^1(0, T; L_2(\Gamma)) \to L_2(0, T; X), \tag{7.51}$$

where X is a Hilbert space satisfying $L_2(\Omega) \times H^{-1}(\Omega) \subset X$. To show (7.51), we first note that if $u \in H_0^1(0, T; L_2(\Gamma))$, then by problem (7.12) with $f = 0$, we obtain

$$\frac{d^2 \mathcal{L}u}{dt^2} = \mathcal{A}\mathcal{L}u \in L_2(0, T; H^{-\frac{3}{2}-\varepsilon}(\Omega) \times H^{-2}(\Omega)) \tag{7.52}$$

upon applying [L-M.1, p. 85] to the regularity (7.51). Thus, application of the intermediate derivative theorem [L-M.1, p. 15] between (7.50) and (7.52) yields (7.51) with $X = H^{-\frac{1}{2}-\varepsilon/2}(\Omega) \times H^{-1}(\Omega)$, as desired.

Verification of Assumption (A.2)

Proposition 7.6. The family of operators $K[\eta]$ defined by (7.25) satisfies both

(a) the assumption of collective compactness (7.32), (7.33) on the space $\mathcal{E}_T = L_2(0, T; L_2(\Omega) \times H^{-1}(\Omega))$;

(b) the assumption of strong convergence (7.34).

Proof.

(a) From (7.25) and (7.48) we readily obtain via (4.29),

$$(K[\eta]g)(t) = \int_0^t e^{\mathcal{A}(t-\tau)} \mathcal{F}'[\eta]g(\tau)d\tau = \left| \begin{matrix} \int_0^t S(t-\tau)f'(\eta_1(\cdot))g_1(\tau, \cdot)d\tau \\ \int_0^t C(t-\tau)f'(\eta_1(\cdot))g_1(\tau, \cdot)d\tau \end{matrix} \right| \quad (7.53)$$

$$: \text{continuous } L_1(0, T; L_2(\Omega)) \to C([0, T]; L_2(\Omega) \times H^{-1}(\Omega))$$

uniformly in $\eta \in \mathcal{E}_T$, $\qquad\qquad\qquad\qquad\qquad (7.54)$

where uniformity is a consequence of assumption (7.13). Moreover

$$\left(\frac{dK[\eta]g}{dt} \right)(t) = \left| \begin{matrix} \int_0^t C(t-\tau)f'(\eta_1(\cdot))g_1(\tau, \cdot)d\tau \\ f'(\eta_1(\cdot))g_1(t, \cdot) - A \int_0^t S(t-\tau)f'(\eta_1(\cdot))g_1(\tau, \cdot)d\tau \end{matrix} \right| \quad (7.55)$$

$$: \text{continuous } L_1(0, T; L_2(\Omega)) \to C([0, T]; L_2(\Omega) \times H^{-1}(\Omega))$$

uniformly in $\eta \in \mathcal{E}_T$. $\qquad\qquad\qquad\qquad\qquad (7.56)$

Application of Aubin's Compactness Lemma to (7.54) and (7.56) yields at once that $K[\eta]$ is compact on $\mathcal{E}_T$; i.e., property (A.2)(a$_1$) = (7.32); and indeed, because of the uniform bounds in (7.54) and (7.56), then property (A.2)(a$_2$) = (7.33) on collective compactness attains.

(b) With $\eta, g \in \mathcal{E}_T$, i.e., $\eta_1, g_1 \in L_2(0, T; L_2(\Omega))$, we consider from (7.53)

$$\left(\left| \begin{matrix} A^{\frac{1}{2}} & 0 \\ 0 & A^{\frac{1}{2}} \end{matrix} \right| K[\eta]g \right)(t) = \left| \begin{matrix} \int_0^t A^{\frac{1}{2}} S(t-\tau)f'(\eta_1(\cdot))g_1(\tau, \cdot)d\tau \\ A^{\frac{1}{2}} \int_0^t C(t-\tau)f'(\eta_1(\cdot))g_1(\tau, \cdot)d\tau \end{matrix} \right|. \quad (7.57)$$

Next, by the uniform bound in (7.13) we have $\eta_1 \to f'(\eta_1)$: continuous $L_2(0, T; L_2(\Omega)) \to$ bounded sphere of $L^\infty(\mathbb{R})$. Thus, by Alaoglou's Theorem, there exists a sequence $\eta_{1n} \in L_2(0, T; L_2(\Omega))$ such that $f'(\eta_{1n})$ converges to some $f_0 \in L^\infty(R)$ weak star. Define the operator $\mathcal{F}_0 \in \mathcal{L}(Y)$ by

$$\mathcal{F}_0 y = \left| \begin{matrix} 0 \\ f_0(\cdot)y_1(\cdot) \end{matrix} \right| \quad (7.58)$$

for $y \in [y_1, y_2] \in Y$. Then we have

$$\left| \begin{matrix} A^{\frac{1}{2}} & 0 \\ 0 & A^{\frac{1}{2}} \end{matrix} \right| K[\eta_n] \to \left| \begin{matrix} A^{\frac{1}{2}} & 0 \\ 0 & A^{\frac{1}{2}} \end{matrix} \right| R\mathcal{F}_0 \text{ weakly in } \mathcal{E}_T, \quad (7.59)$$

as it follows by the Lebesgue dominated convergence theorem using the uniform bound in (7.13), weak star convergence and the bound $\|C(t)\| + \|A^{\frac{1}{2}} S(t)\| \leq$const, in the $L_2(\Omega)$-uniform norm. Moreover, in a similar way, one obtains

$$\frac{dK[\eta_n]}{dt} \to \frac{dR\mathcal{F}_0}{dt} \text{ weakly in } \mathcal{E}_T, \quad (7.60)$$

since from (7.53),

$$\left(\frac{dK[\eta_n]}{dt} g \right)(t) = \left| \begin{matrix} \int_0^t C(t-\tau)f'(\eta_{1n}(\cdot))g_1(\tau, \cdot)d\tau \\ f'(\eta_{1n}(\cdot))g_1(t, \cdot) - A^{\frac{1}{2}} \int_0^t A^{\frac{1}{2}} S(t-\tau)f'(\eta_{1n}(\cdot))g_1(\tau, \cdot)d\tau \end{matrix} \right|. \quad (7.61)$$

As a consequence of the weak convergence of (7.59) and (7.60) and of the compactness of $A^{-\frac{1}{2}}$ on $L_2(\Omega)$, we deduce that, as desired,

$$K[\eta_n] \to R\mathcal{F}_0 \text{ strongly in } \mathcal{E}_T. \tag{7.62}$$

Verification of Assumption (A.3)

Proposition 7.7. The family of operators $K_T[\eta]$ defined by (7.26) satisfies (7.35) as well as the assumption of weak convergence (7.36).

Proof. From (7.26) and (7.48) we readily obtain via (4.29)

$$K_T[\eta]g = \int_0^T e^{A(T-t)}\mathcal{F}'[\eta]g(t)dt = \left| \begin{matrix} \int_0^T S(T-t)f'(\eta_1(\cdot))g_1(t,\cdot)dt \\ \int_0^T C(T-t)f'(\eta_1(\cdot))g_1(t,\cdot)dt \end{matrix} \right| \tag{7.63}$$

$$\text{continuous } L_1(0,T;L_2(\Omega)) \to H = H_0^1(\Omega) \times L_2(\Omega), \tag{7.64}$$

and (7.35) is verified. To show (7.36), one proceeds through an argument similar to the one in part (b) of the preceding Proposition 7.6.. As in (7.59), we obtain

$$\left| \begin{matrix} A^{\frac{1}{2}} & 0 \\ 0 & A^{\frac{1}{2}} \end{matrix} \right| K_T[\eta_n] \to \left| \begin{matrix} A^{\frac{1}{2}} & 0 \\ 0 & A^{\frac{1}{2}} \end{matrix} \right| R_T\mathcal{F}_0, \text{ weakly on } L_2(\Omega) \times [\mathcal{D}(A^{\frac{1}{2}})]', \tag{7.65}$$

which is equivalent to $K_T[\eta_n] \to R_T\mathcal{F}_0$, weakly in $\mathcal{D}(A^{\frac{1}{2}}) \times L_2(\Omega)$.

Verification of (A.5) and (A.6)

Proposition 7.7. We have

(i)　　　trajectory $\{(R\mathcal{F}(0))(t), 0 < t < T\} \subset$ compact set of $\mathcal{E}_T =$　(7.66)
　　　　$L_2(0,T;L_2(\Omega) \times H^{-1}(\Omega))$;

(ii)　　　$R_T\mathcal{F}(0) \in H = \mathcal{D}(A^{\frac{1}{2}}) \times L_2(\Omega)$.

Proof. (i) From (7.23), (7.48),

$$(R\mathcal{F}(0))(t) = \int_0^t e^{A(t-\tau)}\mathcal{F}(0)d\tau = \left| \begin{matrix} \int_0^t S(t-\tau)\{f(0)\}(\cdot)d\tau \\ \int_0^t C(t-\tau)\{f(0)\}(\cdot)d\tau \end{matrix} \right| \in C([0,T];\mathcal{D}(A^{\frac{1}{2}})\times L_2(\Omega)), \tag{7.67}$$

and part (i) follows. Part (ii) is then contained in (7.67) with $t = T$.

Verification of assumption (C.1)

For $\gamma = 1$ in the statement of Theorem 7.3, the content of assumption (C.1) is the following result of exact controllability of the linear system corresponding to (7.12), which requires no geometrical conditions if $\Gamma_0 = \emptyset$, $\Gamma_1 = \Gamma$.

Theorem 7.8. Let $f \equiv 0$ in (7.12), and let $T > 0$ be sufficiently large. Then for any given pair $\{w_0, w_1\} \in H_0^1(\Omega) \times L_2(\Omega)$, there exists a suitable control function $u \in$

$H_0^1(0, T; L_2(\Gamma))$ such that the corresponding solution of problem (7.12) with $f \equiv 0$ satisfies

$$w(T, \cdot) = w_t(T, \cdot) = 0 \quad \text{and} \quad \{w, w_t\} \in C([0, T]; H^{\frac{1}{2}}(\Omega) \times L_2(\Omega)). \tag{7.68}$$

By time reversibility, the origin $\{0, 0\}$ can be steered to all of $H_0^1(\Omega) \times L_2(\Omega)$ at time $t = T$, by using $H_0^1(0, T; L_2(\Gamma))$-control functions. Thus, for such T:

$$\mathcal{L}_T : \text{continuous } H_0^1(0, T; L_2(\Gamma)) \quad \text{onto} \quad H_0^1(\Omega) \times L_2(\Omega). \tag{7.69}$$

For a proof of this result, see [L-T.17, Appendix A] (but the regularity in (7.68) was noted before in (2.25)). A different approach is given in [Lio.3].

Verification of assumption (C.2)

Verification of conditions (7.43) and (7.44) of assumption (C.2) (appropriate controllability of the linearized problem and its limit version in the sense of (7.34) and (7.36)) amounts to the same 'uniqueness property' of the corresponding homogeneous problem, as explained below. To verify (7.43), we consider the problem

$$\begin{cases} \zeta_{tt} = \Delta\zeta + f'(\eta_1)\zeta & \text{in } (0, T] \times \Omega = Q & \text{(7.70a)} \\ \zeta|_{t=0} = \zeta_t|_{t=0} = 0 & \text{in } \Omega; & \text{(7.70b)} \\ \zeta|_\Sigma = u & \text{in } (0, T] \times \Gamma = \Sigma, & \text{(7.70c)} \end{cases}$$

with η_1 a fixed element of $L_2(0, T; L_2(\Omega))$ which corresponds to the linearized abstract version of (7.14) with $\mathcal{F}(0) = 0$. If $M_T[\eta]$ is the map defined in (7.40) we have

$$M_T[\eta] : u \to \{\zeta(T), \zeta_t(T)\} : \mathcal{U}_T = H_0^1(0, T; L_2(\Gamma)) \to H = \mathcal{D}(A^{\frac{1}{2}}) \times L_2(\Omega).$$

Let ϕ be the solution to the corresponding homogeneous problem backward in time:

$$\begin{cases} \phi_{tt} = \Delta\phi + f'(\eta_1)\phi; & \text{(7.71a)} \\ \phi|_{t=T} = \phi_0 = y_1 \in L_2(\Omega); \quad \phi_t|_{t=T} = \Phi_1 = -Ay_0 \in H^{-1}(\Omega); & \text{(7.71b)} \\ \phi|_\Sigma \equiv 0. & \text{(7.71c)} \end{cases}$$

Then, by multiplying problem (7.70) and problem (7.71) by ζ and integrating by parts, we obtain

$$\frac{d^2}{dt^2} M_T^*[\eta] \begin{vmatrix} y_0 \\ y_1 \end{vmatrix} = \frac{\partial\phi}{\partial\nu}(t; \phi_0, \phi_1); \quad \phi_0 = y_1, \quad \phi_1 = -Ay_0. \tag{7.72}$$

To test the injectivity condition (7.43) on M_T^* of assumption (C.2), we let

$$0 \equiv M_T^*[\eta] \begin{vmatrix} y_0 \\ y_1 \end{vmatrix}, \quad [y_0, y_1] \in H; \quad \eta \in \mathcal{E}_T, \tag{7.73}$$

and we want to deduce that in fact $[y_0, y_1] = 0$. Now, (7.73) implies by (7.72),

$$\frac{d^2 M_T^*[\eta]}{dt^2} \begin{vmatrix} y_0 \\ y_1 \end{vmatrix} = \frac{\partial\phi}{\partial\nu}|_\Sigma \equiv 0$$

for the solution ϕ of (7.71). Thus, to verify (7.43) and (7.44) of (C.2), we need the following uniqueness result, with $p = f'(\eta_1)$, $\eta_1 \in L_2(0, T; L_2(\Omega))$ for (7.43) and, respectively, $p = f_0$, with f_0 any of the $L_\infty(Q)$-functions obtained as a limit above (7.58) for (7.44).

Theorem 7.9. Consider the problem

$$
\begin{cases}
\phi_{tt} = \Delta\phi + p(t, x)\phi & \text{in } Q: & (7.74a) \\
\phi|_{t=0} = \phi_0 \in L_2(\Omega); \phi_t|_{t=0} = \phi_1 \in H^{-1}(\Omega); & \text{in } \Omega; & (7.71b) \\
\phi|_\Sigma = \frac{\partial\phi}{\partial\nu}|_\Sigma \equiv 0 & \text{in } \Sigma, & (7.71c)
\end{cases}
$$

with $p \in L_\infty(Q)$. Let $T > 0$ be sufficiently large. Then, in fact, $\phi_0 = \phi_1 = 0$.

Remark 7.3. (i) The above is, apparently, a new uniqueness result due to the fact that $p(t, x)$ is only L_∞ in t and x. A proof of Theorem 7.9 is given in [L-T.17]: it uses three multipliers $h \cdot \nabla\phi$, ϕ and ϕ_t, $h(x) = x - x_0$, in the style of recent exact controllability results, to boost the original regularity of the initial data $\{\phi_0, \phi_1\}$ from $L_2(\Omega) \times H^{-1}(\Omega)$ to $H_0^1(\Omega) \times l_2(\Omega)$; at this level, it then applies the uniqueness result of [Hor.1] to conclude that $\phi_0 = \phi_1 = 0$.

Having verified (A1)–(A.5), and (C.1), (C.2), we conclude that Theorem 7.3 for $\gamma = 1$ is a specialization of the abstract Theorem 7.4.

The case $\gamma = 0$. In the case of the statement of Theorem 7.3 with $\gamma = 0$, we select the following spaces for the abstract framework of Theorem 7.4:

$$
Y = L_2(\Omega) \times H^{-1}(\Omega); \qquad H = L_2(\Omega) \times H^{-1}(\Omega); \tag{7.75}
$$

$$
U = L_2(\Gamma); \qquad \mathcal{U}_T = L_2(0, T; L_2(\Gamma)); \tag{7.76}
$$

$$
\mathcal{E}_T = L_2(0, T; L_2(\Omega) \times H^{-1}(\Omega)). \tag{7.77}
$$

All the necessary assumptions are verified as in the case $\gamma = 1$ (the case of assumption (C.2) is simpler now, since now the initial data $\{\phi_0, \phi_1\}$ of problem (7.71) are smoother, i.e., they are in $H_0^1(\Omega) \times L_2(\Omega)$), with the only exception of assumption (A.1): now, in fact, we verify alternative (7.30) (rather than alternative (7.29) as before for $\gamma = 1$). First, the continuity requirement (7.28) for $\mathcal{L}$ with $\mathcal{U}_T$ and $\mathcal{E}_T$ as in (7.75), (7.76) is *a fortiori* true from Section 2,Theorem 2.1. Next, recalling (7.48), we see that assumption (7.30) on R_T is *a fortiori* satisfied provided that the operator R_T in (7.24),

$$
R_T \begin{vmatrix} 0 \\ g_2 \end{vmatrix} = \int_0^T e^{A(T-t)} \begin{vmatrix} 0 \\ g_2(t) \end{vmatrix} dt = \begin{vmatrix} \int_0^T S(T-t)g_2(t)dt \\ \int_0^T C(T-t)g_2(t)dt \end{vmatrix} \tag{7.78}
$$

is compact: $g_2 \in L_1(0, T; L_2(\Omega)) \rightarrow H = L_2(\Omega) \times H^{-1}(\Omega)$, which is certainly true, since, in fact, $R_T \begin{vmatrix} 0 \\ g_2 \end{vmatrix} \in H_0^1(\Omega) \times L_2(\Omega)$.

This way, the abstract assumptions of Theorem 7.4 are verified also for the case $\gamma = 0$ of Theorem 7.3. The cases $0 < \gamma < 1$ then follow by interpolation applied to the operator $(\mathcal{L}_T^*)^{-1}$ as in [L-T.18].

Notes. *The abstract framework of Theorem 7.4 is suitable also for plate problems. A case where all the required assumptions (A.1)–(A.5) and (C.1), (C.2) have been verified is the Euler-Bernoulli equation $w_{tt} + \Delta^2 w = f(w)$ in Q with controls in $w|_\Sigma$ and $\Delta w|_\Sigma$ and I.C. $\{w_0, w_1\} \in [H^2(\Omega) \cap H_0^1(\Omega)] \times L_2(\Omega)$ under the assumption $|f'(r)| + |f''(r)| \leq c, r \in R$. The most delicate assumption to verify — like in the hyperbolic problem (7.12) — is the approximate controllability assumption*

(C.2) *requiring a uniqueness result with irregular time-dependent 'potential' function (like p(t, x)
in (7.74a)). If controls are now put in $w|_\Sigma$ and $\frac{\partial w}{\partial \nu}|_\Sigma$, all assumptions have been verified (in suitable
function spaces) except (C.2). This and other plate problems require uniqueness results which
apparently are not available in the literature at present.*

8. Riccati Operator Equations and Hyperbolic Mixed Problems

8.1. Introduction; the Case where $\mathcal{B}$ and $\mathcal{R}$ are Bounded

The optimal control theory which minimizes a quadratic functional cost over a time
interval $[0, T]$ associated with a linear dynamical equation leads naturally to Riccati
equations: differential if $T < \infty$, and algebraic if $T = \infty$. In fact, the solution of
the Riccati equation provides the pointwise (in t) link between the optimal state and
the control which implements the optimal strategy, in what is generally referred to as
"optimal pair in feedback form" (see e.g., Eqs. (8.4), (8.8), and (8.23) below). Here, we
shall introduce the subject with reference to an abstract equation. We shall then first recall
a few main highlights of the theory in the case where the control operator ($\mathcal{B}$ below), as
well as the observation operator ($\mathcal{R}$ below) are bounded. Next, we shall provide suitable
extensions to unbounded $\mathcal{B}$ and $\mathcal{R}$ (in both cases, $T < \infty$ and $T = \infty$), under some
'abstract' assumptions. These assumptions are particularly tailored toward hyperbolic
problems (as well as plate equations, not necessarily hyperbolic), which motivate these
assumptions in the first place. In later subsections, we shall in fact verify that these
abstract assumptions either coincide with, or else properly contain, special *features* or
properties of hyperbolic (or plate) equations. We should hasten to point out that this
verification is not a routine step: it is crucially based on the recent regularity properties
of hyperbolic problems recalled in Sections 2 and 3 (and of plate equations) as well
as on related recent controllability/uniform stabilization properties for these dynamics
if $T = \infty$, also established only recently. Consider the following abstract differential
equation

$$\dot{y} = Ay + Bu \text{ on, say, } [\mathcal{D}(A^*)]'; \quad y(0) = y_0 \in Y, \tag{8.1}$$

where Y is a Hilbert (state) space, and where A is the generator of a s.c. semigroup on
Y, and if $R(\cdot, A$ is the resolvent of A,

$$\mathcal{B} \in \mathcal{L}(U; [\mathcal{D}(A^*)]'); \text{ equivalently } R(\lambda_0, A)\mathcal{B} \in \mathcal{L}(U;Y), \tag{8.2}$$

where U is another Hilbert (control) space. Here, $[\mathcal{D}(A^*)]'$ is the Hilbert space dual to the
space $\mathcal{D}(A^*) \subset Y$ with respect to the Y-topology, where A^* is the Y-adjoint of A. The
optimal quadratic cost problem associated with (8.1) is: for a preassigned $0 < T \le \infty$,
minimize

$$J(u, y) = \int_0^T \{|\mathcal{R}y(t)|_Z^2 + |u(t)|_U^2\}dt \tag{8.3}$$

over all $u \in L_2(0, T;U)$, where $y(t) = y(t; y_0)$ is the solution of (8.1) due to u. We
shall initially review a few key points of the theory when the operators $\mathcal{B} \in \mathcal{L}(U;Y)$ and
$\mathcal{R} \in \mathcal{L}(Y;Z)$ are bounded.

Case where $\mathcal{B}$ and $\mathcal{R}$ are bounded. [Bal.1] Here the main results are as follows.

(i) Case $T < \infty$. There exists an optimal pair $\{u^0(t), y^0(t)\}$, where $u^0(t) = u^0(t; y_0)$, $y^0(t) = y^0(t; y_0)$, unique solution of the optimal control problem (8.3), such that

$$u^0(t; y_0) = -\mathcal{B}^*\mathcal{P}(t)y^0(t; y_0), \quad 0 \le t \le T, \tag{8.4}$$

where the operator

$$0 \le \mathcal{P}(t) = \mathcal{P}^*(t) : \text{ continuous } Y \to C([0, T]; Y) \tag{8.5}$$

is a solution of the following Differential Riccati Equation for $x, y \in \mathcal{D}(\mathcal{A})$ in the Y-inner product;

$$\begin{cases} \frac{d}{dt}(\mathcal{P}(t)x, y) = -(\mathcal{R}^*\mathcal{R}x, y) - (\mathcal{P}(t)\mathcal{A}x, y) - (\mathcal{A}^*\mathcal{P}(t)x, y) + (\mathcal{P}(t)\mathcal{B}\mathcal{B}^*\mathcal{P}(t)x, y), \\ \quad 0 \le t < T; \\ \mathcal{P}(T) = 0. \end{cases}$$
$$\tag{8.6}$$

Moreover, there is a unique positive definite solution.

(ii) Case $T = \infty$. (Existence) We assume, additionally, the following Finite Cost Condition:

$$\left.\begin{array}{l} \text{for each } y_0 \in Y, \text{ there exists } \bar{u} \in L_2(0, \infty; U)\text{such} \\ \text{that if } \bar{y} = y(\bar{u})\text{is the corresponding solution} \\ \text{of (8.1), then } J(\bar{(u)}, \bar{y}) < \infty. \end{array}\right\} \tag{8.7}$$

Then, there exists a unique optimal pair $\{u^0(t), y^0(t)\}$, $u^0(t) = u^0(t; y_0)$, $y^0(t) = y^0(t; y_0)$, of the optimal control problem (8.3) with $T = \infty$ such that

$$u^0(t; y_0) = -\mathcal{B}^*\mathcal{P}y^0(t; y_0), \quad 0 \le t \le \infty, \tag{8.8}$$

where the operator

$$0 \le \mathcal{P} = \mathcal{P}^* \in \mathcal{L}(Y) \tag{8.9}$$

is a solution of the following Algebraic Riccati Equation for $x, y \in \mathcal{D}(\mathcal{A})$ in the Y-inner product:

$$(\mathcal{A}^*\mathcal{P}x, y) + (\mathcal{P}\mathcal{A}x, y) + (\mathcal{R}^*\mathcal{R}x, y) - (\mathcal{P}\mathcal{B}\mathcal{B}^*\mathcal{P}x, y) = 0. \tag{8.10}$$

(Uniqueness) Assume, in addition, that

there exists an operator $\mathcal{K} \in \mathcal{L}(Y; U)$ such that the s.c. semigroup generated by $\mathcal{A} + \mathcal{K}\mathcal{R}$ is uniformly stable: there exist $C \ge 1$, $\rho > 0$ such that

$$\|e^{(\mathcal{A}+\mathcal{K}\mathcal{R})t}\|_{\mathcal{L}(Y)} \le C e^{-\rho t}, \quad t \ge 0. \tag{8.11}$$

(This is, in particular, the case if $\mathcal{R} > 0$ in which case it suffices to take $\mathcal{K} = -k^2\mathcal{R}^{-1}$ for k^2 large enough.) Then, in fact, the s.c. semigroup $e^{\mathcal{A}_P t}$ on Y, $t \ge 0$, generated by

$$\mathcal{A}_P = \mathcal{A} - \mathcal{B}\mathcal{B}^*\mathcal{P} \tag{8.12}$$

is uniformly stable: there exist $M \ge 1$, $\delta > 0$ such that

$$\|e^{\mathcal{A}_P t}\|_{\mathcal{L}(Y)} \le M e^{-\delta t}, \quad t \ge 0. \tag{8.13}$$

Moreover, the ARE (8.10) admits a unique positive self-adjoint solution.

8.2. The Unbounded Case: Main statements for $T < \infty$

Orientation. In the present section we turn our attention to the optimal control problem (8.3) for the dynamics (8.1), when $\mathcal{B}$ is unbounded and subject to assumption (8.2). Both cases $T < \infty$ and $T = \infty$ will be treated. The analysis of the Riccati equations for an abstract dynamics (8.1), with $\mathcal{A}$ and $\mathcal{B}$ unbounded, strongly depends on the character of the equation modelled by (8.1). It has been recognized [L-T.19] that in order to extract best results, it is imperative to distinguish between different classes. Our emphasis in this paper is on the class of bounded control problems for hyperbolic equations; thus, our treatment here will be motivated by, and ultimately directed to, such class. As it happens, the theory will likewise apply also to 'plate equations' not necessarily hyperbolic, but there is no space in this paper for an adequate inclusion of plate problems. (Also, no further mention will be made of the class where $\mathcal{A}$ generates a s.c., *analytic* semigroup: here, a rather complete and comprehensive theory is available when $\mathcal{A}^{-\gamma}\mathcal{B} \in \mathcal{L}(U; Y)$ for some $0 \le \gamma < 1$, which covers relevant boundary/point control problem for parabolic equations, as well as for plate equations with a strong degree of damping which exhibit a parabolic-like behavior; see [C-T.1], [L-T.19]).

As mentioned in the introduction, our abstract assumptions given below are tuned to hyperbolic (or plate) equations: they will either capture, or else properly contain, intrinsic *properties* of these dynamics. This will be demonstrated in the next Sections 8.3 and 8.4, where we shall make crucial use of the regularity properties of Sections 2 and 3.

8.2.1. The Unbounded Case for $T < \infty$. The Differential Riccati Equation. Theorems 8.1, 8.2

In addition to the standing hypothesis (8.2), we now assume:

$$\mathcal{R} \in \mathcal{L}(\mathcal{D}(\mathcal{A}); Z); \quad \text{equivalently, } \mathcal{R}\mathcal{A}^{-1} \in \mathcal{L}(Y; Z) \qquad (8.14)$$

($\mathcal{D}(\mathcal{A})$ endowed with norm $|y|_{\mathcal{D}(\mathcal{A})} = |\mathcal{A}y|_Y$, $\mathcal{A}$ boundedly invertible without loss of generality since $T < \infty$) and, moreover:

(H.1): the map $\mathcal{R} e^{\mathcal{A}t}\mathcal{B}$ can be extended as a map $\mathcal{R}e^{\mathcal{A}t}\mathcal{B}$: continuous $U \to L_1(0, T; Z)$; i.e.,

$$\int_0^T |\mathcal{R}e^{\mathcal{A}t}\mathcal{B}u|_Z dt \le c_T |u|_U, \quad u \in U. \qquad (8.15)$$

(H.2): the map $\mathcal{R}e^{\mathcal{A}t}$ can be extended as a map $\mathcal{R}e^{\mathcal{A}t}$: continuous $Y \to L_\infty(0, T; Z)$; i.e.,

$$\sup_{0 \le t \le T} |\mathcal{R} e^{\mathcal{A}t}x|_Z \le c_T |x|_Y, \quad x \in Y \qquad (8.16)$$

(H.3.): the map $\mathcal{B}^* e^{\mathcal{A}^* t}\mathcal{R}^*$ can be extended as a map $\mathcal{B}^* e^{\mathcal{A}^* t}\mathcal{R}^*$: continuous $Z \to L_2(0, T; U)$; i.e.

$$\int_0^T |\mathcal{B}e^{\mathcal{A}^* t}\mathcal{R}^* z|_U^2 \, dt \le c_T |z|_Z^2, \quad z \in Z. \qquad (8.17)$$

Theorem 8.1. [L-T.20] Assume hypotheses (H.1) = (8.15), (H.2) = (8.16), (H.3) = (8.17) (in addition to (8.2)). Then there exists a solution $\mathcal{P}(t) \in \mathcal{L}(Y)$, $0 \le t \le T$, of the Differential Riccati Equation

$$\frac{d}{dt}(\mathcal{P}(t)x, y)_Y = -(\mathcal{R}x, \mathcal{R}y)_Z - \mathcal{P}(t)x, Ay)_Y - (\mathcal{P}(t)Ax, y)_Y + (B^*\mathcal{P}(t)x, B^*\mathcal{P}(t)y)_U,$$
$$\forall\, x, y \in \mathcal{D}(A);$$
$$\mathcal{P}(T) = 0$$
$$\tag{8.18}$$

(which, moreover, is given constructively in the proof, in terms of the data of the problem). Such solution $\mathcal{P}(t)$ is the unique solution to enjoy the following properties:

(i) $\mathcal{P}(t) = \mathcal{P}^*(t)$, $0 \le t \le T$ (* in Y); $\qquad\qquad\qquad\qquad\qquad\qquad\qquad$ (8.19)

(ii) $\mathcal{P}(t)$: continuous $Y \to C([0, T]; Y)$; $\qquad\qquad\qquad\qquad\qquad\qquad$ (8.20)

(iii) $B^*\mathcal{P}(t)$: continuous $A \to C([0, T]; U))$; $\qquad\qquad\qquad\qquad\qquad$ (8.21)

(iv) $B^*\mathcal{P}(\cdot)e^{A(\cdot-s)}B$: continuous $U \to L_2(s, T; U)$ for any s, $0 \le s < T$, with norm which may be taken as independent of s; $\qquad\qquad\qquad$ (8.22)

(v)   the optimal pair is related by the pointwise expression
$$u^0(t; y_0) = -B^*\mathcal{P}(t)y^0(t; y_0); \qquad\qquad\qquad\qquad\qquad\qquad (8.23)$$

(vi)   the optimal cost is

$$J(u^0(\cdot, y_0), y^0(\cdot; y_0)) = (\mathcal{P}(0)y_0, y_0)_Y. \quad \square \qquad\qquad (8.24)$$

A more regular case is given by the following result. To state it, we shall postulate the existence of a Hilbert space $\mathcal{U}_{[0,T]} \subset L_2(0, T; U)$ (algebraically and topologically), with restriction $\mathcal{U}_{[t,T]} \subset L_2(t, T; U)$, such that the following assumptions hold true:

(H.4): the operator $\mathcal{L}_0$, defined by

$$(\mathcal{L}_0 u)(\tau) = \int_0^\tau e^{A(\tau-r)}Bu(r)dr \qquad\qquad\qquad\qquad (8.25)$$

satisfies

$$\mathcal{L}_0 : \text{ continuous } \mathcal{U}_{[0,T]} \to C([0, T]; Y). \qquad\qquad\qquad (8.26)$$

(H.5): the map $\mathcal{L}_0^*\mathcal{R}^*\mathcal{R}e^{At}$ can be extended as a map

$$\mathcal{L}_0^*\mathcal{R}^*\mathcal{R}e^{At} : \text{ continuous } Y \to \mathcal{U}_{[0,T]}, \qquad\qquad\qquad (8.27)$$

where

$$(\mathcal{L}_0^* v)(\tau) = \int_\tau^T B^* e^{A^*(r-\tau)}v(r)dr, \quad 0 \le \tau \le T. \qquad\qquad (8.28)$$

(H.6): for each $t \in [0, T]$, $T < \infty$, the map

$$\mathcal{L}_t^*\mathcal{R}^*\mathcal{R}\mathcal{L}_t \text{ is compact } \mathcal{U}_{[t,T]} \to \text{ itself}, \qquad\qquad\qquad (8.29)$$

where

$$(\mathcal{L}_t u)(\tau) = \int_t^\tau e^{A(\tau-r)}Bu(r)dr; \quad (\mathcal{L}_t^* v)(\tau) = \int_\tau^T B^* e^{A^*(r-\tau)}v(r)dr, \quad t \le \tau \le T.$$
$$\tag{8.30}$$

Remark 8.1. Assumption (H.5)=(8.27) is stronger than assumptions (H.2) = (8.16) and (H.3) = (8.17) combined (see [L-T.20, Remark 6.2]). $\quad \square$

In the next theorem we shall give regularity results for the optimal pair $\{u^0(\cdot, t; y_0), y^0(\cdot, t; y_0)\}$ of the optimal control problem as in (8.3), except that integration is now over the interval $[t, T]$ so that the process initiates at the time t from the initial point $y_0 \in Y$.

Theorem 8.2.
(i) Assume hypotheses (H.5) = (8.26) and (H.6) = (8.27). Then

$$\sup_{0 \leq t \leq T} |u^0(\cdot, t; x)|_{\mathcal{U}_{[t,T]}} \leq c_T |x|_Y. \tag{8.31}$$

(ii) Assume, in addition, hypothesis (H.4) = (8.25). Then

$$\sup_{0 \leq t \leq T} |y^0(\cdot, t; x)|_{C([t,T];Y)} \leq c_T |x|_Y. \tag{8.32}$$

Comments on the assumptions. In the subsequent Section 8.3, we shall show that all assumptions (H.1) through (H.6) hold true in the physically relevant case of the boundary control/boundary observation cost functional for hyperbolic mixed problems of Neumann type. In particular, we shall show that assumptions (H.1), (H.3), and (H.6) are verified to hold true in this case in a *crucial* way as a result of the sharp regularity theory reported in Section 3, while earlier theory in instead inadequate and insufficient for this purpose. The following observation may enlighten the assumptions. In Theorem 8.1, both the observation operator $\mathcal{R}$ from Y to Z and the input-solution operator $\mathcal{L}_0$ in (8.25) from $L_2(0, T; U)$ to $L_2(0, T; Y)$ are allowed to be *unbounded*; however, their composition $\mathcal{R}\mathcal{L}_0$ is instead *continuous*: $L_2(0, T; U) \to L_2(0, T; Z)$; i.e., it is 'nicer' than each of its components viewed separately. In the case of the boundary control/boundary observation cost functional for hyperbolic mixed problem of Neumann type on $\dim\Omega \geq 2$ — to be discussed in Section 8.4 below — where the operator $\mathcal{R}$ gives the Dirichlet trace, all this says that the Dirichlet trace of the hyperbolic solution behaves more regularly than it could be inferred from looking at the interior regularity of the solution and applying trace theory (even formally). This property is, in fact, a distinctive feature of wave (and plate) problems, as documented in Sections 2 and 3 in the former case. In case of the hyperbolic problem of Neumann type, we shall see in Section 8.3 that the more regular result in Theorem 8.2 applies with $\mathcal{U}_{[0,T]} = H^{\frac{1}{2}}(\Sigma) = L_2(0, T; H^{\frac{1}{2}}(\Gamma)) \cap H^{\frac{1}{2}}(0, T; L_2(\Gamma))$.

8.3. Application of Theorems 8.1 and 8.2: The DRE for Boundary Control and Boundary Observation for Hyperbolic Mixed Problems of Neumann Type

With Ω an open bounded domain in R^n, $n \geq 2$, with sufficiently smooth boundary Γ, we return to the mixed problem (3.1) of Neumann type, which we rewrite here in a specialized form

$$w_{tt} - \Delta w + w = 0 \qquad \text{in } \Omega = (0, T] \times \Omega; \tag{8.33a}$$

$$w(0, \cdot) = w_0; \ w_t(0, \cdot) = w_1 \quad \text{in } \Omega; \tag{8.33b}$$

$$\frac{\partial w}{\partial \nu} = u \qquad \text{in } \Sigma = (0, T] \times \Gamma; \tag{8.33c}$$

(As noted in Remark 3.1(v), the case $\dim\Omega = 1$ is much more regular, and this drastically simplifies the analysis below.) We consider the optimal control problem: with $0 < T < \infty$

preassigned, minimize

$$J(u, w) = \int_0^T \{|w(t)|_{L_2(\Gamma)}^2 + |u|_{L_2(\Gamma)}^2\}dt \tag{8.34}$$

over all $u \in L_2(0, T; L_2(\Gamma)) = L_2(\Sigma)$, with w solution of (8.33) due to u. We shall show that this optimal control problem is a specialization of problem (8.1) through (8.3).

Abstract setting. The abstract setting for the mixed problem (8.33) is provided in Section 4.2. Let the abstract spaces Y and U of model (8.1), and the observation space Z be

$$Y = H^1(\Omega) \times L_2(\Omega); \quad U = L_2(\Gamma); \quad Z = L_2(\Gamma). \tag{8.35}$$

The operators $\mathcal{A}$ and $\mathcal{B}$ of model (8.1) are given by (4.72) and (4.73) respectively. Finally, the observation operator $\mathcal{R}$ is defined by

$$\mathcal{R} : Y \supset \mathcal{D}(\mathcal{R}) \to Z = L_2(\Gamma) : \mathcal{R}\begin{vmatrix} y_1 \\ y_2 \end{vmatrix} = y_1|_\Gamma = N^* A y_1, \tag{8.36}$$

see (4.62). Thus, $\mathcal{R}$ is certainly defined on $\mathcal{D}(\mathcal{A})$. For $[y_1, y_2] \in \mathcal{D}(\mathcal{R})$ and $z \in Z = L_2(\Gamma)$,

$$\left(\mathcal{R}\begin{vmatrix} y_1 \\ y_2 \end{vmatrix}, z\right)_{L_2(\Gamma)} = (N^* A y_1, z)_{L_2(\Gamma)} = (y_1, Nz)_{\mathcal{D}(A^{\frac{1}{2}})} = \left(\begin{vmatrix} y_1 \\ y_2 \end{vmatrix}, \mathcal{R}^* z\right)_{\mathcal{D}(A^{\frac{1}{2}}) \times L_2(\Omega) = Y} \tag{8.37}$$

Thus

$$\mathcal{R}^* z = \begin{vmatrix} Nz \\ 0 \end{vmatrix}, \quad z \in Z = L_2(\Gamma). \tag{8.38}$$

We now verify assumptions (H.1), (H.2), (H.3) of Theorem 8.1.

Verification of (H.1) = (8.15)

Recalling (4.75), (4.73), and (8.36), we obtain

$$\mathcal{R}e^{\mathcal{A}t}\mathcal{B}u = \mathcal{R}e^{\mathcal{A}t}\begin{vmatrix} 0 \\ ANu \end{vmatrix} = S(t)ANu|_\Gamma. \tag{8.39}$$

Moreover, recalling (4.61) we have

$$AN : \text{continuous } L_2(\Gamma) \to [\mathcal{D}(A^{\frac{1}{4}+\rho})]' = [H^{\frac{1}{2}+2\rho}(\Omega)]'. \tag{8.40}$$

Using Corollary 3.5 with $1 - \theta = \frac{1}{2} + 2\rho$, $w_0 = 0$, and $w_1 = ANu$, we obtain via (4.66) with $f \equiv 0$, $u \equiv 0$:

$$\mathcal{R}e^{\mathcal{A}t}\mathcal{B}u = S(t)ANu|_\Sigma \in H^{\alpha - \frac{1}{2} - 2\rho + (\beta - \alpha)(\frac{1}{2} - 2\rho)}(\Sigma) \tag{8.41}$$

$$\subset L_2(\Sigma) \tag{8.42}$$

continuously, where the last inclusion in the step from (8.41) to (8.42) attains in view of the sharp regularity theory where $\beta \geq \alpha > \frac{1}{2}$. Thus, from (8.42) we obtain

$$|\mathcal{R}e^{\mathcal{A}t}\mathcal{B}u|_{L_2(0,T;L_2(\Gamma))} \leq c_T |u|_{L_2(\Gamma)} \tag{8.43}$$

and assumption (H.1) = (8.15) is *a fortiori* verified with $Z = U = L_2(\Gamma)$ as dictated by (8.35). Note, on the other hand, that the earlier regularity theory where $\alpha = \beta = \frac{1}{2}$ as recalled in Remark 3.3 would not be sufficient to obtain (H.1) = (8.15) from (8.41).

Verification of (H.2) = (8.16)

Let $x = [x_1, x_2] \in Y = H^1(\Omega) \times L_2(\Omega)$. Then by (8.36) and (4.79) we have continuously,

$$\mathcal{R}e^{\mathcal{A}t}x = [C(t)x_1 + S(t)x_2]|_\Gamma \in C([0, T]; H^{\frac{1}{2}}(\Gamma)), \tag{8.44}$$

by direct application of trace theory. Thus

$$|\mathcal{R}e^{\mathcal{A}t}x|_{C([0,T];H^{\frac{1}{2}}(\Gamma))} \leq c_T|x|_{H^1(\Omega) \times L_2(\Omega)}, \tag{8.45}$$

and assumption (H.2) = (8.16) is *a fortiori* verified. Note here that only classical theory, rather than the recent sharp theory as in Theorem 3.3 (in Theorem 4.4) has been used.

Verification of (H.3) = (8.17)

Let $z \in Z = L_2(\Gamma)$. Then by (8.38), in $\mathcal{R}^*$, (4.79) with $\mathcal{A}$ replaced by $\mathcal{A}^* = -\mathcal{A}$ (in our canonical example (8.33); but this is not essential), and (4.82) on $\mathcal{B}^*$, we obtain

$$\mathcal{B}^* e^{\mathcal{A}^*t} \mathcal{R}^* z = \mathcal{B}^* e^{\mathcal{A}^*t} \begin{vmatrix} Nz \\ 0 \end{vmatrix} = \mathcal{B}^* \begin{vmatrix} C(t)Nz \\ AS(t)Nz \end{vmatrix} = AS(t)Nz|_\Gamma. \tag{8.46}$$

By (8.41) with $u = z \in L_2(\Gamma)$, and again *using crucially the sharp regularity theory* as in going from (8.41) to (8.42), we obtain likewise from (8.46),

$$|\mathcal{B}^* e^{\mathcal{A}^*t} \mathcal{R}^* z|_{L_2(0,T;L_2(\Gamma))} \leq c_T|z|_{L_2(\Gamma)}, \tag{8.47}$$

a result that the earlier theory with $\alpha = \beta = \frac{1}{2}$ cannot produce. Thus assumption (H.3) = (8.17) is again verified with $U = Z = L_2(\Gamma)$.

Remark 8.2. With Reference to problem (8.33) with, say, $w_0 = w_1 = 0$, we recall from Remark 3.1(iii) that the operator $\mathcal{L}_0 : u \to \{w, w_t\}$ is *not* continuous $L_2(\Gamma) \to C([0, T]; H^1(\Omega) \times L_2(\Omega))$, unless $\dim \Omega = 1$.

Having verified assumption (H.1), (H.2), (H.3) for the optimal control problem (8.33), (8.34), we conclude that Theorem 8.1 is applicable to it. However, even more is true. We shall now verify that assumptions (H.4) through (H.6) are also fulfilled in this case (recall from Remark 8.1 that assumption (H.5) = (8.27) is stronger than assumptions (H.2) = (8.16) and (H.3) = (8.17) combined).

Thus Theorem 8.2 is likewise applicable to the optimal control problem (8.33), (8.34). We shall then write explicitly the specialization of these two abstract theorems as applied to problem (8.33), (8.34).

Verification of (H.4) = (8.26)

We begin by selecting

$$\mathcal{U}_{[t,T]} = H^{\frac{1}{2}}(\Sigma_t) = L_2(t, T; H^{\frac{1}{2}}(\Gamma)) \cap H^{\frac{1}{2}}(t, T; L_2(\Gamma)) \tag{8.48}$$

with $\Sigma_t = [t, T] \times \Gamma$. Next, we recall that for problem (8.33), with $w_0 = w_1 = 0$, the regularity result (see Theorem 3.7)

$$u \in L_2(0, T; H^{\frac{1}{2}}(\Gamma)) \rightarrow \{w, w_t\} \in C([0, T]; H^1(\Omega) \times L_2(\Omega)) \qquad (8.49)$$

is contained in [M.1], while a stronger result is reported in [L-T.5, Eq. (1.22)]. In any case (H.4), as specified by (8.26) via (8.48), is *a fortiori* verified, and (8.49) is sufficient for this purpose.

Verification of (H.5) = (8.27)

Here, as above, it will suffice to invoke the weaker regularity results of [M.1], without invoking the stronger results of [L-T.5]. Thus, a conservative analysis will suffice. If $x = [x_1, x_2] \in Y = H^1(\Omega) \times L_2(\Omega)$, then as in (8.44), we have

$$\mathcal{R}e^{At}x = [C(t)x_1 + S(t)x_2]|_\Gamma = \psi(t; \psi_0, \psi_1)|_\Gamma \in H^{\frac{1}{2}}(\Sigma_0), \qquad (8.50)$$

$\psi_0 = x_1$, $\psi_1 = x_2$, since plainly $\psi \in H^1(Q)$, and trace theory yields then (8.50). Next, by (8.38) and (8.50),

$$h(t) \equiv \mathcal{R}^*\mathcal{R}e^{At}x = \begin{vmatrix} N\mathcal{R}e^{At}x \\ 0 \end{vmatrix} = \begin{vmatrix} N(\psi(t)|_\Gamma) \\ 0 \end{vmatrix}, \qquad (8.51)$$

and by (8.25) along with (4.82) on $\mathcal{B}^*$ and (4.79) on e^{At}, we obtain

$$\{\mathcal{L}_0^*[\mathcal{R}^*\mathcal{R}e^{A\cdot}x]\}(\tau) = \mathcal{B}^* \int_\tau^T e^{A^*(\sigma-\tau)}h(\tau)d\tau = \left[\int_\tau^T AS(\sigma-\tau)N(\psi(\sigma)|_\Gamma)\right]_\Gamma = \eta(\tau)|_\Gamma, \qquad (8.52)$$

where

$$\begin{cases} \eta_{tt} = \Delta\eta - \eta & \text{in } Q; & (8.53a) \\ \eta|_{t=T} = \eta_t|_{t=T} = 0 & \text{in } \Omega; & (8.53b) \\ \frac{\partial\eta}{\partial\nu}|_\Sigma = \psi|_\Gamma & \text{in } \Sigma. & (8.53c) \end{cases}$$

But $\psi|_\Gamma \in H^{\frac{1}{2}}(\Sigma_0)$ by (8.50) and then the result of [M.1] recalled above gives $\eta|_\Sigma \in H^{\frac{1}{2}}(\Sigma_0)$ for problem (8.53). Thus by (8.52),

$$\mathcal{L}_0^*[\mathcal{R}^*\mathcal{R}e^{A\cdot}] : \text{ continuous } Y = H^1(\Omega) \times L_2(\Omega) \rightarrow \mathcal{U}_{[0,T]} = H^{\frac{1}{2}}(\Sigma_0) \qquad (8.54)$$

as desired, and (H.5) is verified.

Verification of (H.6) = (8.29)

It is at the level of verifying the compactness assumption (8.29) that we shall make crucial use of the sharp regularity results of Section 3; while, in contrast, the earlier results as in [L-M.1], [M.1] are inadequate. By (8.38) on $\mathcal{R}^*$, (8.36) on $\mathcal{R}$ and (8.25), we obtain with $u \in H^{\frac{1}{2}}(\Sigma_0)$:

$$\mathcal{R}^*\mathcal{R}\mathcal{L}_0 u = \begin{vmatrix} N\mathcal{R}\mathcal{L}_0 u \\ 0 \end{vmatrix} = \begin{vmatrix} N(w|_\Gamma) \\ 0 \end{vmatrix} \qquad (8.55)$$

with $w(t) = w(t; 0, 0)$ solution of problem (8.33) with $w_0 = w_1 = 0$. Thus, as in (8.52), we obtain now via (8.55):

$$\{\mathcal{L}_0^* \mathcal{R}^* \mathcal{R} \mathcal{L}_0 u\}(\tau) \qquad = B^* \int_\tau^T e^{A^*(\sigma - \tau)} (\mathcal{R}^* \mathcal{R} \mathcal{L}_0 u)(\sigma) d\sigma$$

$$= \left[\int_\tau^T AS(\sigma - \tau) N_{(w(\sigma)|_\Gamma)} d\sigma \right]_\Gamma = \zeta(\tau)|_\Gamma, \qquad (8.56)$$

where

$$\begin{cases} \zeta_{tt} = \Delta \zeta - \zeta & \text{in } Q; & (8.57a) \\ \zeta|_{t=T} = \zeta_t|_{t=T} = 0 & \text{in } \Omega; & (8.57b) \\ \frac{\partial \zeta}{\partial \nu}|_\Sigma = w|_\Sigma & \text{in } \Sigma. & (8.57c) \end{cases}$$

But, by [M.1], with reference to (8.33) with $w_0 = w_1 = 0$, we have as before

$$u \in H^{\frac{1}{2}}(\Sigma) \to w|_\Sigma \in H^{\frac{1}{2}}(\Sigma) \qquad (8.58)$$

continuously; and in fact by interpolating with $\theta = \frac{1}{2}$ between the implication (3.3) $\to$ (3.6) and (3.10) $\to$ (3.11) (where the C.C. does not count for $\theta = \frac{1}{2}$), we obtain the following *stronger* result

$$u \in H^{\frac{1}{2}}(\Sigma) \to w|_\Sigma \in H^{2\alpha - \frac{1}{2}}(\Sigma) \qquad (8.59)$$

continuously, where α is defined by (3.2). At this stage, we may still use the conservative regularity (8.58) for $w(\sigma)|_\Gamma$ which enters (8.56); however, it is at the level of analyzing $\mathcal{L}_0^* \mathcal{R}^* \mathcal{R} \mathcal{L}_0 u$ in (8.56) that we crucially use the counterpart of result (8.59). i.e.,

$$w|_\Sigma \in H^{\frac{1}{2}}(\Sigma) \to \mathcal{L}_0^* \mathcal{R}^* \mathcal{R} \mathcal{L}_0 u \in H^{2\alpha - \frac{1}{2}}(\Sigma) \qquad (8.60)$$

as applied to problem (8.57). But $2\alpha - \frac{1}{2} > \frac{1}{2}$, see (3.2); so that

$$\text{the injection } H^{2\alpha - \frac{1}{2}}(\Sigma) \to H^{\frac{1}{2}}(\Sigma) \text{ is compact.} \qquad (8.61)$$

Putting together (8.58)–(8.61), we conclude that

$$\mathcal{L}_0^* \mathcal{R}^* \mathcal{R} \mathcal{L}_0 : \text{ continuous } H^{\frac{1}{2}}(\Sigma) \to H^{2\alpha - \frac{1}{2}}(\Sigma); \qquad (8.62)$$

$$: \text{ compact } H^{\frac{1}{2}}(\Sigma) \to H^{\frac{1}{2}}(\Sigma). \qquad (8.63)$$

The analysis leading to the compactness property (8.63) can be repeated with $\mathcal{L}_0$, $\mathcal{L}_0^*$ replaced by $\mathcal{L}_t$, $\mathcal{L}_t^*$ so that (H.6) = (8.29) is likewise verified. Indeed, the family $\mathcal{L}_t^* \mathcal{R}^* \mathcal{R} \mathcal{L}_t$ is collectively compact on $H^{\frac{1}{2}}(\Sigma) = \mathcal{U}_{[0,T]}$. We note, in contrast, that if one has only the theory of [M.1] available (where then $\alpha = \frac{1}{2}$ in (8.59)), the required compactness of the map in (8.63) *cannot* be verified.

Conclusion

In view of the verification of assumptions (H.1) through (H.6) we conclude that both Theorems 8.1 and 8.2 apply to the optimal control problem (8.33), (8.34). By specializing these theorems, we then obtain

Theorem 8.3. With reference to the optimal control problem (8.34) for the Neumann hyperbolic dynamics (8.33) the following hold true.

(i) There exists a unique solution of the Differential Riccati Equation

$$\frac{d}{dt}(\mathcal{P}(t)x, y)_{H^1(\Omega) \times L_2(\Omega)} = -(x_1|_\Gamma, y_2|_\Gamma)_{L_2(\Gamma)} - (\mathcal{P}(t)x, \mathcal{A}y)_{H^1(\Omega) \times L_2(\Omega)}$$
$$- (\mathcal{P}(t)\mathcal{A}x, y)_{H^1(\Omega) \times L_2(\Omega)} + [\mathcal{P}(t)x]_2|_\Gamma, ([\mathcal{P}(t)y]_2)_{L_2(\Gamma)}$$
$$\forall \, x, y \in \mathcal{D}(\mathcal{A}) = \mathcal{D}(A) \times H^1(\Omega) \tag{8.64}$$

with $\mathcal{P}(T) = 0$, where we write $\mathcal{P}(t)x = \{[\mathcal{P}(t)x]_1, [\mathcal{P}(t)x]_2\}$ for the two components in $H^1(\Omega) \times L_2(\Omega)$.

Uniqueness is within the class of the following properties:

(i_1) $\qquad\qquad \mathcal{P}(t) = \mathcal{P}^*(t) \geq 0. \quad 0 \leq t \leq T \; (* \text{ in } H^1(\Omega) \times L_2(\Omega));$ $\qquad$ (8.65)

(i_2) $\qquad \mathcal{P}(t): \text{ continuous } H^1(\Omega) \times L_2(\Omega) \to C([0, T] : H^1(\Omega) \times L_2(\Omega));$ $\qquad$ (8.66)

(i_3) $\qquad\qquad |\mathcal{P}(\;)x]_2|_\Gamma|_{C([0,T];L_2(\Gamma))} \leq c_T \, |x|_{H^1(\Omega) \times L_2(\Omega)} \cdot$ $\qquad$ (8.67)

(ii) Moreover, the pointwise feedback representation of the optimal pair is

$$u^0\left(t; \left|\begin{matrix} w_0 \\ w_1 \end{matrix}\right|\right) = -\left[\mathcal{P}(t)\left|\begin{matrix} w^0(t; w_0, w_1) \\ w^0_t(t, w_0, w_1) \end{matrix}\right|\right]_2\Big|_\Gamma. \tag{8.68}$$

(iii) The optimal cost is

$$J\left(u^0\left(\cdot, \left|\begin{matrix} w_0 \\ w_1 \end{matrix}\right|\right), \left|\begin{matrix} w^0(\cdot; w_0, w_1) \\ w^0_t(\cdot, w_0, w_1) \end{matrix}\right|\right) = \left(\mathcal{P}(0)\left|\begin{matrix} w_0 \\ w_1 \end{matrix}\right|, \left|\begin{matrix} w_0 \\ w_1 \end{matrix}\right|\right)_{H^1(\Omega) \times L_2(\Omega)} \tag{8.69}$$

(iv) The optimal control satisfies

$$\sup_{0 \leq t \leq T} |u^0(\cdot, t; x)|_{H^{\frac{1}{2}}(\Sigma_t)} \leq c_T |x|_{H^1(\Omega) \times L_2(\Omega)}. \tag{8.70}$$

(v) The optimal solution satisfies

$$\sup_{0 \leq t \leq T} |y^0(\cdot, t; x)|_{C([t,T];H^1(\Omega) \times L_2(\Omega))} \leq c_T |x|_{H^1(\Omega) \times L_2(\Omega)}. \tag{8.71}$$

8.4. The Unbounded Case for $T = \infty$. The Algebraic Riccati Equation. Theorem 8.4.

In this section we state an existence and uniqueness theorem for solvability of the Algebraic Riccati Equation associated with the optimal control problem (8.3) with $T = \infty$, for the abstract dynamics (8.1) subject to a certain assumption. Applications will include second-order hyperbolic mixed problems of Dirichlet type to be analyzed in Section 8.5 (as well as many other plate equations, not necessarily hyperbolic; but there is no space in this paper to include plate problems; we can only refer to [L-T.19], [L-T.24].

Theorem 8.4. [L-T.8], [L-T.21], [F-L-T] (Existence) With reference to problem (8.3) with $T = \infty$ for the dynamics (8.1), assume that

(A.1) the map $\mathcal{B}^* e^{\mathcal{A}^* t}$ can be extended as a map $\mathcal{B}^* e^{\mathcal{A}^* t}$: continuous $Y \to L_2(0, T; U)$; i.e.,

$$\int_0^T |\mathcal{B}^* e^{\mathcal{A}^* t} x|_U^2 \leq c_T |x|_Y^2; \tag{8.72}$$

(A.2) the observation operator $\mathcal{R}$ is bounded:

$$\mathcal{R} \in \mathcal{L}(Y, Z); \tag{8.73}$$

(A.3) the Finite Cost Condition (8.7) holds true.

Then there exists a self-adjoint, non-negative solution

$$0 \leq \mathcal{P} = \mathcal{P}^* \in \mathcal{L}(Y) \tag{8.74}$$

of the Algebraic Riccati Equation

$$(\mathcal{P}x, \mathcal{A}y)_Y + (\mathcal{P}\mathcal{A}x, y)_Y + (\mathcal{R}x, \mathcal{R}y)_Z - (\mathcal{B}^*\mathcal{P}x, \mathcal{B}^*\mathcal{P}y)_U = 0, \quad \forall x, y \in \mathcal{D}(\mathcal{A}) \tag{8.75}$$

such that:

(i) $$\mathcal{P} \in \mathcal{L}(\mathcal{D}(\mathcal{A}); \mathcal{D}(\mathcal{A}_p^*)) \cap \mathcal{L}(\mathcal{D}(\mathcal{A}_p), \mathcal{D}(\mathcal{A}^*))$$

where the operator

$$\mathcal{A}_p = \mathcal{A} - \mathcal{B}\mathcal{B}^*\mathcal{P} \tag{8.76}$$

generates a s.c. semigroup on Y; thus, the ARE (8.75) holds true also for all $x, y \in \mathcal{D}(\mathcal{A}_p)$;

(ii) $$\mathcal{B}^*\mathcal{P} \in \mathcal{L}(\mathcal{D}(\mathcal{A}); U) \cap \mathcal{L}(\mathcal{D}(\mathcal{A}_p); U); \tag{8.77}$$

(iii) the optimal cost due to the unique optimal pair $u^0(t) = u^0(t; y_0)$, $y^0(t) = y^0(t, y_0)$ is given by

$$J(u^0, y^0) = (\mathcal{P}y_0, y_0)_Y; \tag{8.78}$$

(iv) $$u^0(t; y_0) = -\mathcal{B}^*\mathcal{P}y^0(t; y_0), \tag{8.79}$$

where (8.79) is understood a.e. in t if $y_0 \in Y$; which instead, if $y_0 \in \mathcal{D}(\mathcal{A}_p)$, then (8.77) implies $y^0(t; y_0) \in C([0, T]; \mathcal{D}(\mathcal{A}_p))$ and by (8.79), $u^0(t; y_0) \in C([0, T]; U)$ for any $T > 0$.

(Uniqueness) In addition to the above hypotheses, assume the following (so-called 'detectability' condition (D.C.)):

(D.C.): There exists $\mathcal{K} : Z \supset \mathcal{D}(\mathcal{K}) \to Y$ densely defined such that

$$|\mathcal{K}^*x|_Z \leq c\{|\mathcal{B}^*x|_U + |x|_Y\}, \quad \forall x \in \mathcal{D}(\mathcal{B}^*) \subset Y, \tag{8.80}$$

so that the operator

$$\mathcal{A}_K = \mathcal{A} + \mathcal{K}\mathcal{R} \tag{8.81}$$

once closed (and denoted by the same symbol) is the generator of a s.c. semigroup $e^{\mathcal{A}_K t}$ on Y, which is then assumed to be exponentially stable on Y:

$$|e^{\mathcal{A}_K t}|_{\mathcal{L}(Y)} \leq M_K e^{-\omega_K t}, \quad t > 0, \tag{8.82}$$

for some $M_K \geq 1$; $\omega_K > 0$. (For $\mathcal{R} > 0$, we choose $\mathcal{K} = -c^2\mathcal{R}^{-1}$ with constant c^2 sufficiently large, and the detectability condition (8.80)–(8.82) is automatically satisfied.) Then

(a) the solution $\mathcal{P}$ to the ARE (8.75) is unique within the class of non-negative self-adjoint operators in $\mathcal{L}(Y)$ which satisfy the regularity properties (8.77);

(b)   the s.c. semigroup $e^{A_p t}$ generated by A_p in (8.76) is exponentially, uniformly stable on Y.

Remark 8.3. Note that the above Theorem 8.4 states only that $B^* P$ is densely defined, see (8.77). Indeed, it can be shown that if $\exp At$ is a s.c. *group* uniformly bounded for negative times (the case of all conservative systems like wave and plate equations), then $B^* P$ cannot be bounded unless B is bounded (the less interesting case).

In the next subsection, we shall see that the regularity results of Section 2 for second-order hyperbolic equations of Dirichlet type will enable us to apply Theorem 8.4 and thus obtain existence (and uniqueness) of the Algebraic Riccati Equation which arises in the case of the optimal control problem for wave equations with Dirichlet control.

8.5. Application of Theorem 8.4: The ARE for Boundary Control for Hyperbolic Mixed Problems of Dirichlet Type

We return to probelm (2.1) which we rewrite here

$$\begin{cases} w_{tt} = \Delta w & \text{in } Q; & (8.83a) \\ w(0, \cdot) = w_0; w_t(0, \cdot) = w_1 & \text{in } \Omega; & (8.83b) \\ w|_\Sigma = u & \text{in } \Sigma; & (8.83a) \end{cases}$$

$$\{w_0, w_1\} \in L_2(\Omega) \times H^{-1}(\Omega). \qquad (8.83d)$$

We consider the optimal control problem: minimize

$$J(u, w) = \int_0^t \left\{ |w(t)|^2_{L_2(\Omega)} + |w_t(t)|^2_{H^{-1}(\Omega)} + |u(t)|^2_{L_2(\Gamma)} \right\} dt. \qquad (8.84)$$

We shall show that this optimal control problem is a specialization of problem (8.1) through (8.3), to which Theorem 8.4 is applicable [L-T.8].

Abstract setting. The abstract setting for the mixed problem (8.83) is given by Section 4.1. The spaces are

$$Y = L_2(\Omega) \times H^{-1}(\Omega); \quad Z = Y; \quad U = L_2(\Gamma), \qquad (8.85)$$

and the operators A and B are defined in (4.28), (4.29), respectively, while

$$\mathcal{R} = \text{ Identity on } Y \qquad (8.86)$$

so that (A.2) = (8.73) is trivially true. The standing assumption (8.2) holds true, as it follows plainly from (4.30b).

Verification of assumption (A.1) = (8.72)

This is precisely statement (4.44).

Verification of assumption (A.3) on the Finite Cost Condition

This is a consequence of recent theories of exact controllability of problem (8.83) which show (in particular) that: for $T >$ some $T_0 > 0$ given any $\{w_0, w_1\} \in L_2(\Omega) \times H^{-1}(\Omega)$

there exists $u \in L_2(0, T; L_2(\Gamma))$ such that $w(T) = w_t(T) = 0$; thus setting $u \equiv 0$ for $t > 0$ we obtain that the Finite Cost Condition is satisfied [B-R-L], [L-T.22], [L-T.23], [H.1], [Lio.3], [Tr.5], [Ta.1].

Conclusion. Theorem 8.4 is then applicable to problem (8.83), (8.84) and specializes to the following statement.

Theorem 8.5. With reference to the optimal control problem (8.84) for the Dirichlet hyperbolic dynamics (8.83), the following holds. There exists a unique $\mathcal{P} = \mathcal{P}^* \geq 0$ bounded on Y in (8.85) such that the ARE (8.75) holds true with $\mathcal{A}$ in (4.28), $\mathcal{R}$ in (8.86), and with

$$(\mathcal{B}^*\mathcal{P}x, \mathcal{B}^*\mathcal{P}y)_{L_2(\Gamma)} = \left(\frac{\partial[\mathcal{P}x]_2}{\partial \nu}, \frac{\partial[\mathcal{P}y]_2}{\partial \nu} \right)_{L_2(\Gamma)},$$

where we write $\mathcal{P}x = \{[\mathcal{P}x]_1, [\mathcal{P}x]_2\}$ for the two components of $\mathcal{P}x$ in $L_2(\Omega) \times H^{-1}(\Omega)$. Moreover, the corresponding closed loop problem (8.83) obtained by using

$$u(t) = \frac{\partial}{\partial \nu} \left\{ \mathcal{P} \begin{bmatrix} w(t) \\ w_t(t) \end{bmatrix} \right\}_2 \tag{8.88}$$

in (8.83d), is (exponentially) uniformly stable in $\mathcal{L}(L_2(\Omega) \times H^{-1}(\Omega))$. $\square$

References

[A.1]   H. Amann, Parabolic evolution equations with nonlinear boundary conditions, J. Diff. Eqn. **12** (1988), 201–269

[Bal.1]   A. V. Balakrishnan, Applied Functional Analysis, Springer-Verlag, 1976

[B-B]   P. L. Butzer and H. Berens, Semigroups of Operators and Approximation, Springer-Verlag, 1967

[B-L-R.1] C. Bardos, G. Lebeau and J. Rauch, Sharp sufficient conditions for observation, control and stabilization of waves from the boundary, SIAM J. Control & Optim.

[C.1]   G. Chen, Energy decay estimates and exact controllability for the wave equation in a bounded domain, J. Math. Pures et Appl. **9** (1979), 249–274

[C-L.1]   S. Chang and I. Lasiecka, Riccati equations for nonsymmetric and nondissipative hyperbolic systems with L_2-boundary controls, J. Math. Anal. & Appl. **116** (1986), 378–414

[C-T.1]   S. Chen and R. Triggiani, Proof of extensions of two conjectures on structural damping for elastic systems: The case $\frac{1}{2} \leq \alpha \leq 1$, Pacific J. of Mathematics, **136** (1989), 15-55

[Fa.1]   H. O. Fattorini, Boundary control systems, SIAM J. Control, **6** (1968), 349–385

[Fa.2]   H. O. Fattorini, Second Order Linear Differential Equations in Banach Spaces, North Holland, 1985

[F-L-T.1] F. Flandoli, I. Lasiecka, and R. Triggiani, Algebraic Riccati Equations with non-smoothing observation arising in hyperbolic and Euler-Bernoulli equations, Annali di Matematica Pura e Applicata, (iv) **CLIII** (1988), 307–382

[Fu.1]   D. Fujawera, Concrete characterization of domains of fractional powers of some elliptic operators of second order, Proc. Japan Acad. **43** (1967), 82–86

[G.1]   P. Grisvard, Caracterization de quelques espaces d'interpolation, Arch. Rat. Mech. and Anal. **25** (1967), 40–63

[H.1]   F. L. Ho, Observabilité frontiere de l'equation des ondes, C. R. Acad. Sc. Paris **302** (1986), 443–446

[Hor.1]   L. Hormander, The Analysis of Linear Partial Differential Operators. I, II, III, IV. Springer-Verlag 1983, 1983, 1985, 1985

[K.1]   V. A. Kondratiev, Boundary problems for elliptic equations in domains with conical or angular points, Trans. Moscow Math. Soc. 16 (1967)

[Lag.1]   J. Lagnese, Decay of solutions of wave equations in a bounded region with boundary dissipation, J. Diff. Eqns. **50** (2) (1983), 163–182

[Las.1]   I. Lasiecka, Unified theory for abstract parabolic boundary problems: A semigroup approach, Appl. Math. and Optim. **6**, 4 (1980), 281–333

[Las.2]   I. Lasiecka, Stabilization of the semilinear wave equation with viscous damping, J.Diff. Eqns., **86** (1990), 73-87

[Las.3]   I. Lasiecka, Sharp regularity results for mixed hyperbolic problems of secod order, Lecture Notes in Mathematics 1223, Differential Equations in Banach Spaces, Springer-Verlag (1986), 160–175, Proceedings of a conference held at the University of Bologna, Italy, July 1985

[L-S.1]   I. Lasiecka and A. Stahel, The wave equation with semilinear Neumann boundary conditions, Nonlinear Anal. Methods and Appl. **15** (1990), 39–58

[L-L-T.1] I. Lasiecka, J. L. Lions, and R. Triggiani, Non homogeneous boundary value problems for second order hyperbolic operators, J. Math. Pures et Appl. **69**, 1986, 149–192

[L-T.1]   I. Lasiecka and R. Triggiani, A cosine operator approach to modeling $L_2(0, T; L_2(\Gamma))$-boundary input hyperbolic equations, Appl. Math. and Optim. **7** (1981), 35–83

[L-T.2]   I. Lasiecka and R. Triggiani, Regularity of hyperbolic equations under $L_2(0, T; L_2(\Gamma))$-boundary terms, Appl. Math. and Optim. **10** (1983), 275–286

[L-T.3]   I. Lasiecka and R. Triggiani, Sharp regularity results for mixed second order hyperbolic equations of Neumann type: The L_2-boundary case, Annali di Matem. Pura e Appl., IV **CLVII** (1990), 285-367

[L-T.4]   I. Lasiecka and R. Triggiani, Sharp regularity theory for second order hyperbolic equations of Neumann type, Rendiconti Classe die Scienze fisiche, Matematiche e naturali, Atti della Accademia Nazionale dei Lincei, Roma, Vol. LXXXIII (1989)

[L-T.5]   I. Lasiecka and R. Triggiani, Regularity theory of hyperbolic equations with non-homogeneous Neumann boundary conditions, Part II: General boundary data, J. Diff. Eqns., **94** (1991), 112-164

[L-T.6]   I. Lasiecka and R. Triggiani, Trace regularity of the solutions of the wave equation with homogeneous Neumann boundary conditions and compactly supported data, J. Math. Anal. and Appl., **14** (1989), 49–71

[L-T.7]   I. Lasiecka and R. Triggiani, A lifting theorem for the time regularity of solutions to abstract equations with unbounded operators and applications to hyperbolic equations, Proceedings Amer. Math. Soc. **104** (1988), 745–755

[L-T.8]   I. Lasiecka and R. Triggiani, Riccati equations for hyperbolic partial differential equations with $L_2(0, T; L_2(\Gamma))$-Dirichlet boundary terms, SIAM J. Control & Optimiz. **24** (1986), 884–926

[L-T.9]   I. Lasiecka and R. Triggiani, Dirichlet boundary control problem for parabolic equation with quadratic cost: Analyticity and Riccati's feedback synthesis, SIAM J. Control & Optimiz. **21** (1983), 41–67

[L-T.10] I. Lasiecka and R. Triggiani, The regulator problem for parabolic equations with Dirichlet boundary control. Part I: Riccati's feedback synthesis, and regularity of optimal solutions, Appl. Math. and Optimiz., **16** (1987), 147–168

[L-T.11] I. Lasiecka and R. Triggiani, Hyperbolic equations with Dirichlet boundary feedback via position vector: Regularity and almost periodic stabilization, Parts I, II, III, Appl. Mathem. and Optimiz. **8** (1981), 1–37; **8** (1982), 103–130; **8** (1982), 199–221

[L-T.12] I. Lasiecka and R. Triggiani, Dirichlet boundary stabilization of the wave equation with damping feedback, J. Math. Anal. & Appl. **97** (1983), 112–130

[L-T.13] I. Lasiecka and R. Triggiani, Nondissipative boundary stabilization of hyperbolic equations with boundary observation, J. de Mathem. Pures et Appl. **63** (1984), 59–80

[L-T.14] I. Lasiecka and R. Triggiani, Feedback semigroups and cosine operators for boundary feedback parabolic and hyperbolic equations, J. Diff. Eqns. **47** (1983), 246–272

[L-T.15] I. Lasiecka and R. Triggiani, Structural assignment of Neumann boundary feedback parabolic equations: The case of trace in the feedback loop, Annali Mat. Pura & Appl. (IV) **XXXII** (1982), 131–175

[L-T.16] I. Lasiecka and R. Triggiani, Stabilization and structural assignment of Dirichlet boundary feedback parabolic equations, SIAM J. Control & Optimiz. **21** (1983), 766–803

[L-T.17] I. Lasiecka and R. Triggiani, Exact controllability of semi-linear abstract systems with applications to wave and plate problems, Appl. Mathem. and Optimiz., **23** (1991), 109-154

[L-T.18] I. Lasiecka and R. Triggiani, Exact controllability for the wave equation with Neumann boundary control, Appl. Math. & Optimiz. **19** (1989), 243–290

[L-T.19] I. Lasiecka and R. Triggiani, Algebraic Riccati Equations arising in boundary/point control: A review of theoretical and numerical results, in Perspectives in Control Theory, Proceedings of Sielpia Conference, Sielpia, Poland, 1988; Jakubczyk-Malanowski-Respondek Edits, Birkhäuser, 1990

[L-T.20] I. Lasiecka and R. Triggiani, Differential Riccati Equations with unbounded coefficients: Application to boundary control/boundary observation hyperbolic problems, J. of Non-Linear Analysis **17** (1991), 655-682

[L-T.21] I. Lasiecka and R. Triggiani, Infinite horizon quadratic cost problems for boundary control problems, Proceedings of the 26K Conference on Decision and Control, Los Angeles, California, December 1987, pp. 1005–1010

[L-T.22] I. Lasiecka and R. Triggiani, Uniform exponential energy decay of wave equations in a bounded region with $L_2(0, \infty; L_2(\Gamma))$-feedback control in the Dirichlet boundary conditions, J. Diff. Eqts. **66** (1987), 340-390

[L-T.23] I Lasiecka and R. Triggiani, Uniform stabilization of the wave equation with Dirichlet or Neumann-feedback control without geometrical conditions, Appl. Math. & Optim **25** (1992), 189-224

[L-T.24] I Lasiecka and R. Triggiani, Differential and Algebraic Riccati Equations
with Application to Boundary/Point Control Problems: continuous theory
and Approximation Theory, Springer Verlag Lecture Notes LNCIS vol. **164,**
150 pp., 1991

[Lio.1]   J. L. Lions, Optimal Control of Systems Governed by Partial Differential
Equations, Springer-Verlag (1971) (English translation of French edition by
Dunod and Gauthier-Villar, Paris, 1968)

[Lio.2]   J. L. Lions, Contrôle des systèmes distribues singuliers, Gauthier-Villars, 1983

[Lio.3]   J. L. Lions, Contrôllabilité exacte, perturbations et stabilizations des systèmes
distribues, Masson 1989, Paris.

[L-M.1]   J. L. Lions, and E. Magenes, Nonhomogeneous boundary value problems and
applications I, II (1972) Springer-Verlag

[M.1]   S. Myatake, Mixed problems for hyperbolic equations of second order, J. Math.
Kyoto Univ. 130–3 (1973), 435–487

[N.1]   J. Necas, Les méthodes directes en théorie des equations elliptiques, Masson
Edits, 1967

[Sak.1]   R. Sakamoto, Mixed problems for hyperbolic equations, I, II, J. Math. Kyoto
Univ., **10-2** (1970), 343–347; and **10-3** (1970), 403–417

[Sak.2]   R. Sakamoto, Hyperbolic boundary value problems, Cambridge University
Press (1982)

[Sy.1]   W. W. Symes, A trace theorem for solutions of the wave equation, Math.
Methods in appl. Sciences **5** (1983), 35–93

[St.1]   A. Stahel, Hyperbolic initial boundary value problems with nonlinear boundary
conditions, Nonlinear Analysis **13** (1988), 231–257

[Ta.1]   D. Tateru, Ph.D. dissertation, University of Virginia, May 1993

[Tr.1]   R. Triggiani, A cosine operator approach to modeling $L_2(0, T; L_2(\Gamma))$-
boundary input problems for hyperbolic systems, Lecture Notes CIS Springer-
Verlag (1978), 380–390. Proceedings 8th IFIP Conference, University of
Würzburg, W. Germany, July 1977

[Tr.2]   R. Triggiani, On Nambu's boundary stabilizability problem for diffusion
processes, J. Diff. Eqn. **33** (1979), 189–200. (Preliminary versions in
Proceedings International Conference on Recent Advances in Differential
Equations, Miramare-Trieste (Italy), August 1978), Academic Press, and
Proceedings International Conference on Systems Analysis, I.R.I.A., Paris,
1978)

[Tr.3]   R. Triggiani, Well-posedness and regularity of boundary feedback parabolic
systems, J. Diff. Eqns. **36** (1980), 347–362

[Tr.4]   R. Triggiani, Boundary feedback stabilizability of parabolic equations, Appl.
Math. & Optimiz. **6** (1980), 201–220

[Tr.5]   R. Triggiani, Exact boundary controllability on $L^2(\Omega) \times H^{-1}(\Omega)$ for the wave
equation with Dirichlet control acting on a portion of the boundary and related
problems, Appl. Math. and Opt. **18** (1988), 241–277

[Tr.6]   R. Triggiani, Announcement of sharp regularity theory for second order
hyperbolic equations of Neumann type, Springer-Verlag Lectures Notes,
Proceedings IFIP Conference on Optimal Control of Systems Governed by
Partial Differential Equations, University of Santiago de Compostele, Spain,
July 1987

[W.1]   D. C. Washburn, A semigroup theoretic approach to modeling of input problems, SIAM J. Control **17** (1979), 652–671

[Z.1]   E. Zuazua, Exact boundary controllability for the semilinear wave equation, in "Non-linear PDE and their applications", Research Notes in Math., Pitman (1989)

Springer-Verlag
and the Environment

We at Springer-Verlag firmly believe that an international science publisher has a special obligation to the environment, and our corporate policies consistently reflect this conviction.

We also expect our business partners – paper mills, printers, packaging manufacturers, etc. – to commit themselves to using environmentally friendly materials and production processes.

The paper in this book is made from low- or no-chlorine pulp and is acid free, in conformance with international standards for paper permanency.